Fierce
Climate
Sacred
Ground

Fierce
Climate
Sacred
Ground

An Ethnography of Climate Change
in Shishmaref, Alaska

Elizabeth Marino

University of Alaska Press

University of Alaska Press
P.O. Box 756240
Fairbanks, AK 99775-6240

ISBN: 978-1-60223-266-2 (paperback); ISBN: 978-1-60223-267-9 (electronic)

Library of Congress Cataloging-in-Publication Data

Marino, Elizabeth K., 1978–
    Fierce climate, sacred ground : an ethnography of climate change in Shishmaref,
Alaska / by Elizabeth Marino.
        pages cm
    Includes bibliographical references and index.
    ISBN 978-1-60223-266-2 (pbk.: acid-free paper)—ISBN 978-1-60223-267-9
(electronic)
    1. Human beings—Effect of climate on—Alaska—Shishmaref.   2. Ethnology—
Alaska—Shishmaref.   3. Climatic changes—Alaska—Shishmaref.   4. Indiginous
peoples—Alaska—Shishmaref—Ecology.   5. Shishmaref (Alaska)—Remote-sensing
maps.   I. Title.
    GF71.M37 2015
    304.2'5097986—dc23
                                                                            2015002919

Cover design by Krista West
Interior design by Amnet
Cover images by Elizabeth Marino

This publication was printed on acid-free paper that meets the minimum requirements for ANSI / NISO Z39.48–1992 (R2002) (Permanence of Paper for Printed
Library Materials).

This book is dedicated, in gratitude, to Daniel Iyatunguk, Shirley Weyiouanna and all ancestors from the Shishmaref coast; to Rachel and Rich Stasenko, Esther Iyatunguk, Kate and John Kokeok, Tony and Fannie Weyiouanna, Clifford Weyiouanna, Nancy Kokeok, Annie Weyiouanna, Fred Eningowuk, and all of the adults who make Shishmaref the exquisite homeland it is today; and to Edward, Carter, baby Beth and all the kids from Shishmaref who will inherit the strength and intelligence of their extended families who reach back to time immemorial.

This book is also dedicated to Chris.

All and any of the mistakes in this book are my own.

Contents

———➤•◦•◄———

Figures and Tables   ix

Acknowledgments   xi

Chapter One   It's the End of the World, and Shishmaref
                     Is Everywhere   1

Chapter Two   Unnatural Natural Disasters   19

Chapter Three   Flooding and Erosion in Shishmaref: The Anatomy
                     of a Climate Change Disaster   31

Chapter Four   Seal Oil Lamps and Pre-Fab Housing: A History
                     of Colonialism in Shishmaref   45

Chapter Five   Finding a Way Forward: Trust, Distrust,
                     and Alaska Native Relocation Planning in the
                     Twenty-first Century   61

Chapter Six   The Tenacity of Home   81

Chapter Seven   The Ethics of Climate Change   93

Notes   101

Index   113

Figures and Tables

FIGURES

Figure 1.1   Original map by Emmanuelle Bournay, "Atlas environnement 2007 du Monde diplomatique," Paris. Edited by UA Press for black and white printing.   8

Figure 1.2   Shishmaref is located on the narrow island of Sarichef on the west coast of Alaska. The town is bordered on one side by the ocean-facing Chukchi Sea, and on the other by Shishmaref Inlet. (UA Press)   12

Figure 2.1   Conceptual diagram of vulnerability (Adapted from Adger 2006)   25

Figure 3.1   Bluff erosion in Shishmaref (Photo courtesy of Tony Weyiouanna)   39

Figure 3.2   Weather map of 2011 "Super Storm" (Map National Weather Service "Bering Strait Superstorm," November 9, 2011)   40

Figure 3.3   The end of the seawall and a house in Shishmaref (Photo by Elizabeth Marino)   40

Figure 3.4   Map of predicted and historical shorelines of Shishmaref, Alaska (Alaska District, Corps of Engineers, Civil Works Branch)   42

Figure 4.1   Map by Josh Wisniewski (2011), based on work by Burch (2006)   47

Figure 4.2   Sarichef Island historical development (Map of island taken from Mason et al. in press, radiocarbon dates from Mason 1996)   48

Figure 6.1   Erosion and relocation coalition banner   82

TABLES

Table 3.1   Changes in Weather Patterns   34

Table 3.2   Changes in Permafrost   35

Table 3.3   Changes to Thermokarst Ponds   36

Table 3.4   Changes to Freeze-up and Break-up   37

Table 3.5   Changes in Coastal Erosion Patterns   38

Table 5.1   Interviews on Risk   71

Table 5.2   Interviews on Disaster and Diaspora   72

Table 5.3   Interviews on Communication   73

Acknowledgments

⟫•◦•⟪

I've been lucky in my life to have had tremendous teachers, some of whom I would like to thank here.

I am deeply indebted to Peter Schweitzer, who was a trusted advisor, teacher, and friend throughout my graduate work at the University of Alaska–Fairbanks and the writing of my dissertation—which eventually became this book. Working with Peter allowed me to witness interdisciplinary climate change research that remained authentic to the social complexities and issues of voice, power, and representation so critical to anthropology. I am also thankful for Patricia Kwachka, who was a tireless advocate and a biting critic—at just the right times—and to David Koester and Patrick Plattet, who gave critical feedback and support.

Stacey Paniptchuk was my initial research partner and friend in Shishmaref, and is probably 95 percent of the reason this project came to be. In an extraordinary stroke of luck, she also always makes me laugh. This research also received support from the Shishmaref City Council and the Shishmaref Erosion and Relocation Coalition. I will never be able to thank the community of Shishmaref and these institutions enough. It was my great honor to work with you. Thank you to everyone who consented to be interviewed and the many hours of work and dedication that residents gave to this project.

I am grateful for Christopher Pangle, who edited a late version of this manuscript and is a constant inspiration and demonstration of the writing life.

I remain forever thankful to Roy Agloinga, Rita Buck, Pretty Buck, Ida Lincoln, Joey Simon, Lolene Buck, Peter Buck, and all my friends from White Mountain—who first taught me about the immeasurable beauty of rural Alaska—and I am extremely lucky to have Chanda Meek, Elizabeth Daniels, (the late) Deanna Kingston, (the late) Octavio Sanchez, Steven Affeldt, Mo Hamza, Tony Oliver-Smith, Gregory Button, Heather Lazrus, Julie Maldonado, Amy Harper, Natalie Dollar, Susan Crate, Sveta Yamin-Pasternak, Amber Lincoln, Tom and Lesley Downing, Nicole Peterson, Laura Henry-Stone, Patti Oksotarok Lillie, and Robin Bronen as colleagues, coauthors, writer-friends, and teachers over the years. I am inspired by your work.

This project could not have been completed without support from the National Science Foundation Arctic Social Science Program under the direction of Anna Kerttula de Echave (Grant No. 0713896). I am thankful for all of my colleagues with Moved by the State, including Florian Stammler, Tim Heleniak, Tobias Hozlehner,

Beth Mikow, Gertrude Elimsteiner-Saxinger, Alla Bolotova, Elena Khlinovskaya Rockhill, and Peter Evans. This group of thinkers and academics are the best of the Arctic anthropological tradition, and I am humbled by their rigor and dedication.

I thank my students at Oregon State University–Cascades for all the times they've written more than they thought they could and for their bravery in engaging the challenging and frightening material that has the potential to broaden their visions of the world.

I offer sincere gratitude to Laura Wolfe Lornitzo, Nasuġruk Rainy Higbee, Tina Walker Davis, Neva Hickman, Ryan Hughes, Steve Pool, Beth Simak, Jeni Rinner, Jenni Peskin, Emily Gregg, Jason Turner, Tamera and Mike Ruesse, Ryan McGladrey, Madolyn Orr, Richard Shackell, Shawn Duffy, Amy Mouton, Rhiannon Roberts, and Jessica Hammerman for keeping me interested and engaged with the world.

I am thankful to my mother and stepfather, Judy and Daniel Soulier, and my incomparable brother, Brett Marino, who teach me about family, service, hard work, and vocation; and to my son, Emile Marino, who is the calm center of the universe.

Mostly I thank my daughter, Louise, who doesn't like it when her mama goes to work, but pretends to write books and conduct research when she plays make believe. I love you, kiddo.

This book is dedicated to my best friend and partner, Christopher Wolsko, and my friends and family in Shishmaref, Alaska.

Chapter One

It's the End of the World, and Shishmaref Is Everywhere

The raging sea is tremendously powerful and needs to be respected. Shishmaref will need to be protected from the sea and moved to a different location, in due time. The move needs to be closely tied to our hunting traditional cultural practices. We are sea mammal hunters.
—Herbert Nayokpuk 2005[1]

SHISHMAREF IS EVERYWHERE

The first time I flew into Shishmaref it was 2002. I was a twenty-two-year-old newspaper reporter in a brand-new goose-down jacket holding tight to the seat cushion on a single-engine Alaska bush plane as it bumped through the low layer of clouds. Cold and curious, I stared out the window at a slip of an island that seemed dangerous, vulnerable, and impossibly elegant. The coastal Arctic, especially to an outsider raised on romantic notions of the last frontier, is full of both beauty and ferocity. The Shishmaref island chain is especially stunning. It sits as a curve of sand and permafrost in summer, snow and ice in winter, cutting an arc of differentiation between the Chukchi Sea and the shallow Shishmaref Inlet. Even from that first flight, the wild Arctic coast was magnificent and awe-inspiring, and it seemed to be, from my naïve perspective, part of a Jeffersonian dream of wilderness and frontier freedom.

From the air I could see a string of houses, an airstrip, a haphazard scattering of boats, seal-drying racks, and abandoned machinery. In the relative absence of other signs of human life, these human artifacts took on animated qualities. The houses dug into the tundra and clung to the shore. The boats crept toward the water's edge. The seal racks grew up and out of the ground. Flying into a rural Arctic village felt like flying into a more focused, less abstract world, something I still feel today. Everything in the landscape—human, animal, plant, and plastic—has sharp lines and exists in its own right, with some history.

Of course, at the time I had no idea what I was seeing or why it was organized the way that it was.

The last time I flew into Shishmaref I was a thirty-two-year-old anthropologist, five months pregnant, in the same—now beat to hell—down jacket, happy to be coming home. This time Shishmaref was not an idea or a metaphor. The seal racks were full of real memories. The houses were the homes of Clifford and Shirley, Tony and Fannie, Kate and John, and (most importantly for me) Rich and Rachel. I couldn't wait to eat, and I was craving caribou soup, black meat, and seal oil. The Alaska coast was still menacing, and I looked for changes along the seawall and down the coast from the seawall where there was no protection from the waves. The ocean was also, I knew, part of the social life of Shishmaref. It was a place where you could travel, find food, practice tradition, experience beauty; the ocean was an extension of the village, not distinct from it. The ocean, the land, and the village were part of the same socioecological system that encompassed the coast and everything within it. In the interim ten years since that first flight into Shishmaref, many things in my life had changed, and landing in Shishmaref was less like landing in a different, idealized world and more like going to visit old and dear friends and family.

In the interim ten years, Shishmaref's place in the world had changed as well.

Shishmaref had been a familiar place to me and other Bering Strait residents and researchers for a long time and then, quite suddenly, became a place that had been exported to the world's imagination. Researchers and the popular press had identified Shishmaref as one of the very first victims of human-caused climate change. This ever-growing spotlight of attention focused on one thread of the Shishmaref story. In this narrative, an Iñupiaq[2] hunting community, living a traditional lifestyle for thousands of years, was now vulnerable to catastrophic flooding and loss of traditional homeland because of a fearsome and rapidly changing environment suffering under the effects of climate change. In many ways, this narrative is true.

But the story of Shishmaref is much more complicated than that.

<hr />

Shishmaref is one of a group of communities around the world to experience an increasing number of disasters linked to human-caused, or anthropogenic, climate change. Flooding has become habitual in Shishmaref. This flooding, paired with increasing erosion and loss of habitable land, is forcing the community to consider migration as the only possible response. As landscapes, precipitation patterns, and climates change because of an overall increase in global temperature, some places and some communities that have previously been habitable are likely to become uninhabitable because of fires, floods, erosion, storms, and other natural disasters. Land will literally be lost, in some cases, because of sea-level rise or erosion. In other cases, areas that have historically experienced rare flooding or drought conditions will instead be exposed to habitual flooding and regularly occurring drought, making rebuilding infrastructure cost-prohibitive or creating deserts where land becomes no longer profitable, and people have no viable way of making a living. People may be pushed to move when these conditions occur. Those communities, families, and individuals who move because of extreme changes in climate are known as environmental migrants, or climate refugees, and are an important point of discussion in the debates about climate change among scientists, governments, and the communities themselves.

There has been an explosion of awareness about environmental migrants and environmental migration linked to climate change in both popular and scientific dialogues in recent years. When I teach courses on climate change at Oregon State University–Cascades, Tuvalu, the Maldives, and "those villages in Alaska"—places that have been identified as communities of potential environmental migration linked to climate change—are known and recognized by many of my students. Each month I receive emails from colleagues, friends, and family members with links to stories they've seen about Shishmaref or Alaska and the migrations associated with climate change. I have been interviewed by multiple media outlets preparing stories on Shishmaref (*USA Today*, *The Guardian*, *Financial Times* [UK], the Munich Re Foundation Newsletter) and contacted by other graduate students and faculty asking for assistance, literature reviews, and direction in studying environmental migration linked to climate change in general and climate-change-influenced migration in Alaska specifically.

From my perspective as a researcher who was invested early in the topic of migration in Alaska driven by ecological change, I have witnessed firsthand the crest of interest in and enthusiasm for (1) climate change, (2) migration linked to climate change, and (3) Shishmaref as a quintessential example of these two phenomena. In the summer of 2012, as one of my OSU students was completing a research project on evangelical environmentalism and creation care, he exclaimed during an in-class presentation, "Shishmaref is everywhere!"

To be sure, Shishmaref appears omnipresent from where I sit—in my inbox, in my classroom, in the newspaper stories I read, and in the interviews I conduct—because this is my field, the focal point of my research, and the center point of my attention for the last nine years. But there is something absurd about an outsider's claim that this 600-person, primarily Iñupiaq village in extremely rural West Coast Alaska is "everywhere." The questions for my research, therefore, became (1) what is really happening in Shishmaref, and (2) why is it eliciting so much attention?

SHISHMAREF AND THE GREAT CLIMATE DEBATE

There is little doubt among scientists that the atmospheric temperature is increasing because of greenhouse gas emissions released through human activity. The basic mechanisms of anthropogenic climate change are so scientifically rudimentary that the greenhouse gas effect was first introduced in the scientific research literature in 1898 by a Swedish chemist named Svante Arrhenius. What Arrhenius postulated was that heat enters the earth's system from the sun as shortwave solar radiation. Of the radiation energy that hits the earth's surface, some is absorbed, and some is converted into longwave radiation and reflected back into the atmosphere toward space. A portion of this radiating energy is trapped by greenhouse gas molecules in the atmosphere. Greenhouse gas molecules keep heat within the atmosphere instead of allowing all longwave radiation to escape back into space.

The trapping of energy in the atmosphere instead of reflecting it back into space is what scientists refer to as the "greenhouse" effect. The greenhouse effect is essential for the existence of much of the life on the planet. The average temperature of the earth's surface, at which life as we know it exists, is about fifty-eight degrees Fahrenheit. Without the greenhouse effect, the average temperature would be closer to zero degrees Fahrenheit. So, in spite of the relatively small amounts of greenhouse gasses in the earth's atmosphere, we know they have a significant effect on the climate. Because humans have added increasing amounts of greenhouse gasses into the atmosphere, especially carbon dioxide and especially since the Industrial Revolution, we can expect the temperature of the earth to rise. The long-term biophysical changes to the earth expected as a consequence of rising temperatures linked to greenhouse gas emissions should challenge us all to serious personal and political action regardless of political affiliation, socioeconomic position, or cultural background.

In spite of this mechanistic clarity, as climate change discourses have entered the public sphere, they have proven to be highly contentious. The term "climate change" has come to imply an inexhaustible set of biophysical and ecological phenomena as well as an equally inexhaustible set of values, ethics, personal and political identities, policy recommendations, and agendas. Strategies preventing or limiting extreme climate change—climate change "mitigation" strategies—almost all require a cap on greenhouse gas emissions. Capping emissions means (among other things) limiting the burning of fossil fuels and in most scenarios demands a cost hike for the "cheap" energy that drives many economies.

The "climate change debate" in America quickly polarized as a proxy culture war and became a debate about development versus environmental protection, big business versus big government, and common sense versus organized science. In 1997, the United States Senate passed a resolution to veto any bill that put caps on greenhouse gas emissions and, following this, the United States and Australia refused to ratify the Kyoto Protocol. These decisions set the stage for a political imbroglio that dichotomized climate change camps into "believers versus deniers" and "intervention versus inaction." These competing discourses each sought and continue to seek validation—through both scientific evidence and public consensus.

Involved in this pursuit of verification, scholars and journalists looked for test-case studies to examine how climate change would affect people on the ground. If climate change was as bad as the activists said, then surely someone was being affected in the present. The world needed a face for climate change, and the Arctic appeared to be particularly well-suited for that purpose.

The Arctic experiences something called polar, or Arctic, amplification, which is a greater overall warming in the Arctic region than in other parts of the globe during warming trends. This is due in part to the decrease of snow cover on land and the decrease of sea ice in the Arctic Ocean. Exposed, snow-free land and especially the dark-blue ocean absorb more heat energy than do snow or ice, which reflect light. When the Arctic Ocean in particular is ice free, then the earth's albedo (reflective power) decreases, leading to an increased absorption of light and heat energy. Climate models predict greater warming in the Arctic than in other parts of the world in the coming century because of the increase in heat absorption compared to heat and light reflection. These predictions corroborate recorded evidence of early Arctic warming compared to the rest of the world contemporarily. From 1954 to 2003, the mean annual atmospheric surface temperature in Alaska and Siberia rose between two and three degrees Celsius. This warming has been particularly salient in the winter and spring.[3] Along with warming, snow and ice features have diminished, there has been an increase in windiness[4] and storminess[5] along the coast, and permafrost boundaries have moved north, meaning that previously stable permafrost areas have thawed, causing foundation problems for structures in Alaska and problems with erosion.[6]

What is predicted to happen across the globe is, in clear and measurable ways, occurring in the Arctic right now. Changes in Northern climates, landscapes, and ice features on land and in the ocean have inevitably affected human communities who live in and depend on these ecosystems. Erosion in Alaska Native villages is a primary example of the complex conditions created by a warming environment. Eighty-six percent (184 out of 213) of Alaska Native villages have experienced problems with erosion and flooding.[7]

Erosion in Shishmaref is a primary driver of flooding and infrastructure damage. As permafrost boundaries move north, previously frozen ground in Shishmaref has thawed to become towers of freestanding sand, which deteriorate quickly when exposed to wave action. In 2013 a fall storm removed thirty to forty feet of land in a single night, which is significant for an island that is only a half-mile wide. Added to this, there have been changes in weather patterns. Shore ice, which freezes around the island sometime in the fall, is freezing later and later in the year, meaning that fall storms coming off of the Chukchi Sea no longer meet with a natural ice buffer and instead pummel the island with wind and waves. This erosion and resulting flooding precipitates the need for the community to relocate or migrate. In total there have been six flooding disaster declarations issued for Shishmaref by the State of Alaska since 1988.[8] And so Shishmaref became a community of environmental migrants, or climate refugees, and a face of climate change.

DOES THE ENVIRONMENT MOVE PEOPLE?

The story of environmental migration is not quite as simple as it might seem. To understand how Shishmaref came to be an important climate change case study for researchers and media outlets—to understand how Shishmaref came to be "everywhere"—it is important to understand something about migration itself as a research topic. Throughout the greater part of the twentieth century, social science research on human migration frequently failed to identify natural or environmental systems as driving factors for migration decisions.[9] Some scholars attribute the lack of environmental drivers in human migration research to a Western European/North American bias toward the belief that "technological progress would decrease the influence of nature on human life,"[10] a trend that persisted until well into the latter half of the twentieth century. The idea was that as technology mediated the relationship between people and the environment, the eco-logical niche of any given location was increasingly less important. While, for example, prolonged drought drove Pueblo Grande inhabitants out of the greater Phoenix area during the thirteenth century, the population living there today in prolonged drought will not be forced to move. Instead, the over four million inhabitants will be buffered by technological interventions such as water diversion, air conditioning, and a global mar-ketplace. Within this new, technologically advanced rubric of human-ecological rela-tionships, scholars considered migration to be an economically driven decision, not an environmentally driven one. Poor economies pushed migrants, better economies pulled migrants—the environment was distal as a relevant mechanism for migration.

It was under these intellectual circumstances that a surprising essay by Essam El-Hinnawi, published by the United Nations Environmental Programme in 1985, defined environmental refugees as

> those people who have been forced to leave their traditional habitat, temporarily or permanently, because of a marked environmental disruption (natural and/or triggered by people) that jeopardized their existence and/or seriously affected the quality of their life.[11]

This essay reorganized our modern understanding of people and place. Suddenly, under this new intellectual framing, Phoenix residents, and millions of other people living in resource-scarce or disaster-prone areas, were again conceived of as being vul-nerable to prolonged drought or any other disaster. Humans could be "moved" by the environment—like our ancestors before us.

Next, in 1990, the Intergovernmental Panel on Climate Change (IPCC) made the claim that one of the most significant outcomes of anthropogenic climate change on human populations may be forced migration.[12] In the report, the IPCC stated that, as millions are uprooted by shoreline erosion, coastal flooding, and agricultural dis-ruption, the gravest effects of climate change may be on human migration. In 1993, Norman Myers, an ecologist from Oxford, further linked climate change and migra-tion when he estimated that up to one hundred fifty million people could be forced to migrate due primarily to sea-level rise and desertification by the year 2050.[13] In 2008 and 2011, the International Organization on Migration (IOM) released estimates that projected between two hundred million to one billion potential environmental migrants in the coming century.[14] Thus, in just over a quarter of a century, analysis of human migration scenarios changed from failing to recognize the environment as a significant push factor in migration to estimating that as many as *one out of every nine people on the planet* (1 billion environmental migrants out of 8.9 billion people, the estimated population in 2050[15]) could be an environmental migrant.

These large estimates appeared in peer-reviewed theoretical papers,[16] policy reports,[17] and governmental and nongovernmental organization reports,[18] which identified areas of the world that were vulnerable to small or large changes in climate or environmental conditions that could trigger mass migrations. Images of environmental refugees made their way into the popular press. Visions of large-scale environmental refugees from Bangladesh, sub-Saharan Africa, an increasingly arid Mexico, and other vulnerable countries and populations are panic inducing to wealthy nations. With undocumented migrants in both Europe and the United States already making front-page headlines, environmental migrants are particularly moving as a symbol of turmoil (and xenophobia). The climate-refugee story is a powerful narrative for explaining to a general public the significance and scale of climate change outcomes—and so Shishmaref became a test case for this possibility. If we can't, as a nation, collectively organize to successfully relocate 600 Iñupiat people from the west coast of Alaska, what in the world will we do with New Orleans? What will we do with Bangladesh?

The next step for scientists and journalists trying to understand environmental migration was mapping these areas. Maps of environmental migration "hot spots" were quickly drawn up to give a visual representation of evolving reports. In particular, Emmanuelle Bournay created a map (figure 1.1) for the newspaper *Le Monde Diplomatique* that was based on Norman Myers's 2005 report on environmental migrants and the areas so significantly affected by ecological shifts that migration would ensue. Bournay's map was featured on UNEP's website (though later removed) and circulated widely among scholars and policy makers. It is still featured on the Wikipedia site that explains environmental migration.[19]

It is extraordinary that Shishmaref is labeled on this map, as the only other city to be named explicitly is New Orleans. Labels are more commonly given only to entire regions or countries: the Caribbean, Bangladesh, India, the Mekong River delta, Mexico, Haiti, the Yangtze River (the third largest river in the world), and Central Asia, as examples. One label, the Sahel Belt, with a population of *fifty-eight million people*, is one of the poorest areas of Africa and cyclically experiences extensive famine linked to desertification, land degradation, and socioeconomic structures.[20] Excepting Shishmaref, the labels on this map refer to large populations, in many cases under extreme duress—and large, migrating populations under duress are exactly what policy reports on environmental migration highlight and what drives concern about environmental migration from the IPCC and other policy makers. So, again, why care about Shishmaref?

There are three facts that remain relatively uncontested in the Shishmaref case study. First, there is an extremely high probability that the village of Shishmaref will have to be relocated in the foreseeable future because of continued erosion and habitual flooding.[21] Government reports from 2003 and 2009 identified Shishmaref as one of four villages in Alaska (Shishmaref, Kivalina, Koyukuk, and Newtok) that face imminent threat of disaster related to erosion and flooding, and one of three villages

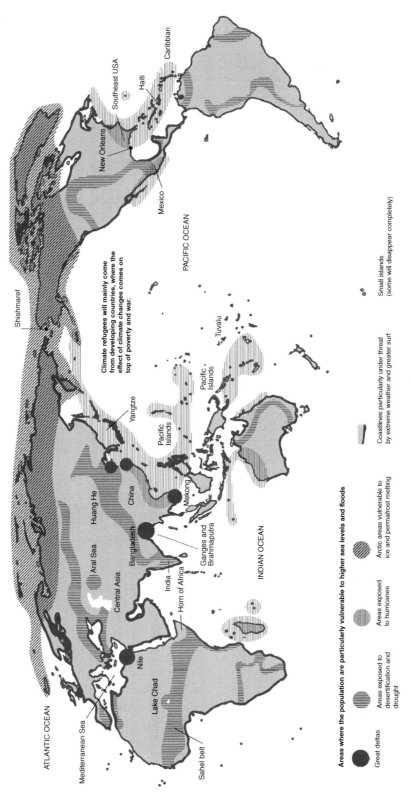

FIGURE 1.1 Original map by Emmanuelle Bournay, "Atlas environnement 2007 du Monde diplomatique," Paris. Edited by UA Press for black and white printing.

(Shishmaref, Kivalina, and Newtok) that would likely need to relocate in the next ten to fifteen years to avoid being "lost to erosion."[22] Second, as discussed above, human-created climate change has already affected Alaska and the Arctic more dramatically and more quickly than other parts of the globe. Third, increased erosion[23] has led to increased vulnerability to flooding[24] for multiple villages in Alaska, and erosion along coastal areas in the North is increasing at greater rates today than in the past, due in part to increasing temperatures.[25]

Whether and how these three uncontested facts are connected in Shishmaref is less clear. However, the combined effect of these three indisputable circumstances— (1) that Shishmaref will have to relocate because of erosion and flooding; (2) that Alaska and the Arctic have had, more than any place on earth, demonstrable effects of anthropogenic global warming, including increased erosion and increased severity of storms; and (3) that increased erosion and flooding events during storms that threaten rural villages are occurring at greater rates today than in previous years—has helped make Shishmaref a quintessential example of environmental migration linked to climate change and has caused researchers and media outlets to focus on this small village as an epicenter of the issue and a meaningful label on the environmental migration map.

In contrast, drought and possible migration linked to drought in the Sahel region has been reoccurring and linked to environmental conditions not directly attributable to anthropogenic climate warming. During the great famine from 1983 to 1985, for example, millions of people in the Sahel belt experienced malnutrition, which caused over four hundred thousand deaths in Ethiopia alone.[26] If environmentally tenuous conditions have always existed in the Sahel Belt, what can we really attribute to climate change? On the other hand, Shishmaref—like the concept of environmental migration itself—burst onto the scene, providing what seemed like an unmistakable example of climate change (the ice is melting!) paired with outcome (the people are fleeing!). Shishmaref became an example of what anthropogenic climate change meant in real terms, for real people, on the ground—and media coverage followed.

Dozens, if not hundreds, of journalists, including those from *Time*, the *New York Times*, and *CNN*, have all done stories on the small village—descending in the summers to talk with people whose houses have fallen into the ocean because of erosion, who have lived through floods and high water, who have hunted seals on thinning ice, and who have considered leaving their ancestral homeland. The Shishmaref story, rightly so, pulls at the heart.

If climate change is the result of industrial activity, then the Iñupiat population in Shishmaref is particularly emotionally salient as a "victim." Iñupiat populations have not traditionally emitted or benefited from large-scale emissions of greenhouse gases. While part of the highly industrialized United States—the largest historical carbon-emitting nation—people in Shishmaref have no cars (there was a single truck on the island when I did fieldwork), mostly have no running water, and were introduced to high-energy-consuming technologies very late in the history of human greenhouse gas emissions. In other words, people in Shishmaref, Alaska, have nearly no culpability for climate change on a large scale and yet are some of the first to suffer the consequences. The least guilty, in this case, are suffering the greatest loss.

The Shishmaref case makes us instinctively consider climate change as a human rights issue. Eskimo hunters on the edge of the world have frequently been used (rightly or wrongly) by Europeans as an example of modernity's opposite, depicted as embodying small-scale, low-tech, and egalitarian ideals. In fact, some of the most famous anthropologists of the last one hundred years built a career around the premise that Iñupiat communities could tell us something about humanity before industrialization. If this Iñupiaq Eskimo community is the clear victim of climate change, a product of modernity and that very industrialization, then something is terribly wrong and unjust.

Yet while Shishmaref was being held out as a paradigmatic victim of climate change, what received less attention in the media was the growing number of scholars who were questioning the climate-refugee narrative. This growing number of researchers considered the reports on potential environmental refugees as "alarmist" in nature and lacking methodological rigor.

Critiques of this "alarmist" literature were numerous. One complaint was that this new research conflated complicated social phenomena into a single driver—environment—and a single outcome—migration. For years, the environment was not considered a significant driver of migration *at all* and now, wholesale and suddenly, climate change researchers were considering it as the *sole* driver of migration. What of economy? What of technology? What of political power? This kind of unilateral simplicity is never how social systems work (which will be discussed more fully in chapter two).

The environmental migration literature also tends to treat all migrations as similar in form and function, when in fact we know that human migration patterns take on a wide variety of forms, including short-term temporary migration, long-term temporary migration, seasonal migration, and age-specific and gender-specific migrations, among others. If a man leaves a community for three months a year for seasonal labor because crops have become less productive in his homeland, is this the same thing as a nationwide migration out of a South Pacific island?

Finally, some scholars pointed out that when social justice and economic intervention occurred in vulnerable areas, the pressure to migrate in most places was removed. If social intervention relieves migration pressure, then is it really the environment that is causing the disaster in the first place?

What is really happening in Shishmaref? Is it fodder for an alarmist story that the news media has perpetuated in order to sell papers and TV commercials? Or is Shishmaref the canary in the coal mine—demonstrating on a small scale the mass migration likely to happen all over the globe because of climate change?

SHISHMAREF, THE FAMILIAR

To fully appreciate what is happening in Shishmaref, Alaska, today, it is important to know something about the land and the people who live there. As an anthropologist,

I looked toward long-term ethnographic engagement as a primary methodology. This research mostly took place in Shishmaref, Alaska—and more specifically took place in the home of Rachel and Rich Stasenko (where I ate, slept, and made friends) and the basement of Shishmaref's Lutheran Church, where there is office space and a public computer. I spent six months in Shishmaref, spread over three trips between 2008 and 2010, conducting ethnographic fieldwork, holding interviews, participating in daily life, playing basketball, putting away black (seal) meat, and conducting a limited survey. This research was done specifically to identify the contributing variables to habitual flooding that, subsequently, drive the need for relocation. I was also interested in the political obstacles to successful relocation.

As the cold and curious journalist in 2002, my personal history with the village, however, reaches back further. In 2002 I was a newspaper reporter for the *Nome Nugget* in the regional hub of Nome, Alaska—the point of departure for flights into the villages of the region. I was assigned to cover the Shishmaref residents' official vote to relocate off of the island. During this trip I witnessed the extreme friendliness of people in Shishmaref and became introduced to the predicament within which the village found itself. In 2005 and 2006, after completing a master's degree, I worked with the US Army Corps of Engineers on the cultural impact assessment of relocating Shishmaref residents to Nome or Kotzebue. This work, again, was captivating and demonstrated the profound local desire to remain in subsistence territory. During this project it became clear to me that protecting the rights of indigenous people to remain in traditional landscapes in the midst of climate change was going to be challenging.

Ethnographic fieldwork is not romantic or easy. I found myself repeatedly embarrassed, incompetent, and lonely in the field. But Shishmaref is an amazing place, full of joy and celebration that is equal or greater to the complex histories of colonialism and the challenges of a changing environment. When I think about Shishmaref now, or when I talk with friends and colleagues from the village, I feel an overwhelming sense of peace, homesickness, and gratitude.

Shishmaref is a small Iñupiaq community in western, coastal Alaska that sits on Sarichef Island just off the coast of the Seward Peninsula between the Shishmaref Inlet and the Chukchi Sea (see figure 1.2).

The population of Shishmaref can fluctuate depending on the time of year, the cost of fuel, and job opportunities. I once asked Tony Weyiouanna, former transportation coordinator for the village, to fact check an academic journal article of mine, and he changed the population estimate from 608 to 609—I always wonder if someone had had a baby! According to the 2000 US Census, the population was 563, and 95 percent of residents on the island are Alaska Native.

Acutely rural, Shishmaref is spatially isolated from the rest of the world, and travel to and from the village to other villages, cities, and states can be challenging. The edge of the world, however, turns out to be a relative concept. If you're a sea mammal hunter, Shishmaref is located in the very center of Iñupiat subsistence hunting practices and

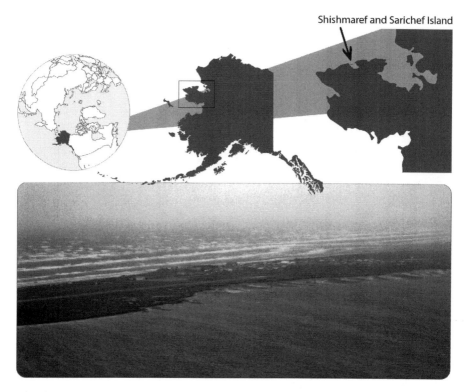

FIGURE 1.2 Shishmaref is located on the narrow island of Sarichef on the west coast of Alaska. The town is bordered on one side by the ocean-facing Chukchi Sea, and on the other by Shishmaref Inlet. (UA Press)

animal migration routes. For many Iñupiat people, the Bering Strait is the center of the world, and Washington, DC—with all its power and prestige—is at the edge.

Still, transportation infrastructure in Iñupiaq territory is scarce. In Shishmaref, small planes and infrequent barges are the only way to transport both goods and people in or out of the village. Travel through traditional hunting and picking territories, over tundra, ice, rivers, and the ocean, is done using all-terrain vehicles (ATVs), snow machines, and small motorboats.

Shishmaref people, the *Kigiqtaamiut* Iñupiat, a linguistic and cultural subgroup of the larger Iñupiaq linguistic group, have inhabited the coastal and river drainage areas around the island for thousands of years, developing a rich tradition and a particular expertise for living in this location. Historically, the *Kigiqtaamiut* food-harvesting techniques and adaptation strategies for the dynamic, animated Alaskan coast have been highly successful in this Arctic landscape.[27] Outsiders often think about traditional Alaskan life being made up of bitter cold, extended darkness, challenging work conditions, and periods of starvation. However, archeological evidence and oral histories suggest that for hundreds of years Iñupiat cultures also experienced life cycles filled with rich storytelling; beautiful, detailed art work; long periods of celebration and rest; and surpluses of food. There were, likewise, periods of in-migration and out-migration, territorial challenges, and political reformations prior to Western influence.

Near the turn of the twentieth century, though, significant changes to traditional (yet dynamic) lifestyles occurred with the discovery of gold in Nome in 1899–1900. With the influx of new residents came the rapid introduction of new technology and Western infrastructure. It was then that a push by missionaries and the state to "civilize" local Native populations began in earnest. Beginning in the early 1900s, the coastal, seasonally nomadic communities that make up today's *Kigiqtaamiut* people began to settle more permanently on Sarichef Island, with the accompanying school, church, and post office.

From 1900 to 1950 most Alaska Native peoples on the Seward Peninsula settled into a smaller village group. This consolidation of population was linked to new infrastructure and legislation that required all school-age children to attend school. In 1959 Alaska became the forty-ninth state in the United States, and Alaska Native peoples entered into a unique and complex sociopolitical organization with federal and state governments through a critical legislative agreement. The Alaska Native Claims Settlement Act (ANCSA), signed into law in 1971, was the biggest land claims act in US history. Through ANCSA, over two hundred village corporations and twelve regional corporations took title of forty-four million acres of land designated for Alaska Native Peoples. The US government additionally paid the twelve corporations $962.5 million to withdraw any right to further land claims. Today Shishmaref is politically organized as a city, a village corporation, and a federally recognized tribe. Residents also belong to the regional corporation, the Bering Strait Native Corporation, and the accompanying nonprofit, Kawerak. Despite these political, technological, and economic changes, there is striking continuity between past and present.

Kigiqtaamiut translates to "people of the island," from the stem noun *Kigiqtaq*, meaning island, and the suffix *miut*, meaning "the people of." *Kigiqtaq* was the proper name of the island where the village now sits, according to friends in Shishmaref. The "island-ness" of Shishmaref still dominates daily life, even as technology and wage-labor jobs make their way into the village. Passage on and off the island via ATV, boat, or snow machine is possible only when the ice is gone (open water) or when the ice is firm (freeze-up). This seasonal ecology creates distinct fall and spring shoulder seasons, during which people stay primarily on the island, and seasons when people easily access the ocean and the mainland to engage in hunting, fishing, and gathering activities, as well as extended camping.

Today, *Kigiqtaamiut* residents continue to harvest and consume an extensive variety of local subsistence foods, including bearded seal, spotted seal, caribou, walrus, musk ox, fish (of all sorts), berries, and greens. The location of the village is uniquely positioned to take advantage of both land and sea mammals; as Fred Eningowuk, a Shishmaref resident, said in an interview, "It's like Shishmaref is in the middle of a circle of subsistence."[28]

The circle metaphor has at least two meanings for Shishmaref residents.

First, food is available throughout the year; as one resource migrates out of traditional hunting and fishing territory, another becomes available. The cyclical turn of the year brings a plentiful variety of subsistence foods, meaning that before the 1900s (and continuing today, though in slightly different ways) hunger was preventable because of the natural patterns of animals, plants, and landscapes. The ice froze and thawed, the seals came and went, the berries ripened, the caribou migrated, the white fish swam under the ice, and the people were fed.

The second metaphorical use of the circle of subsistence is that the people of Shishmaref themselves exist within the natural cycle of the area. I was told more than once that if the people of Shishmaref abandoned the area, the animals would "go away," and the landscape would become fallow. The interdependency of land, animals, and people is profound in long-inhabited landscapes. So in this way, too, Shishmaref people are in the "circle of subsistence," dependent on and depended on by the plant and animal life and the landscapes in which they dwell.

The circle of subsistence is an important feature of the village economy given the limited employment on the island. Nearly 46 percent of adults are not in the cash labor force. Most fixed-income jobs on the island are in government service provision with the tribe, the school, or the local medical clinic. Per capita, the average income in 2010 was $10,203, and almost 27 percent of all residents live below the poverty line.

Limitations of the wage economy in Shishmaref are offset by the subsistence economy. Hunting here is subsidized by part-time employment and/or government transfer payments. Gun shells, gas, nets, motors, and other equipment are purchased with cash and converted into calories through subsistence hunting. These hunts provide residents with healthy, delicious food and link everyone (wage earners, hunters, the women who put away subsistence foods, and the people who are gifted portions of the harvest) to the worldview, lifestyle, moral value system, and sharing practices that make up the *Kigiqtaamiut* culture.

The other important economic sector in Shishmaref surrounds the creation and exportation of Iñupiaq artwork. Shishmaref is the carving and artistic center of the Bering Strait. Small-scale sculptures, jewelry, masks, and carvings of all kinds are produced in Shishmaref by older, experienced carvers and by young, informally apprenticed (mostly) men. These sculptures and jewelry are made of bone, antler, or ivory and are widely distributed throughout the state of Alaska and the world.

Many women in Shishmaref are sewing entrepreneurs, known particularly for their use of spotted sealskin for mittens, hats, slippers, and gloves, and increasingly for their hard-bottom mukluks, though the hard soles are mostly kept within the community. Women also make smaller textiles, which are sold or used as gifts when traveling or when welcoming guests. These include sealskin Christmas tree ornaments and small beaded items. Sewn crafts, such as slippers and gloves, often incorporate beautifully beaded patterns as decoration. One common bead pattern is the Shishmaref star, an intricately beaded star that is unique to the village.

Food and its preparation is another distinguishing aspect of life in the village. Shishmaref is the only village in the Bering Strait region to rely heavily on the bearded seal or *ugruk*. This strong-tasting seal meat is ubiquitous in the village and is kept in white five-gallon buckets stored in many people's *kunituks*, or arctic entryways. The white bucket contains *panaaluk*—dried *ugruk* meat (black meat) that is usually cut thin, with a texture like beef jerky—along with other cuts (e.g., stomach, intestines) of the *ugruk*, and is filled with seal oil. Seal oil is made from rendered seal blubber that varies in opacity. The opacity of the seal oil depends on the quality of blubber, age and type of seal, and conditions under which it was produced. This nutritionally and gastronomically miraculous oil is full of fat and omega-3 fatty acids, which are important for a human diet. Research suggests that daily intake of seal oil significantly lowers glucose intolerance and the preconditions for diabetes[29] and prevents scurvy, the legendary disease suffered by European whalers of the North.

Aside from health benefits and taste, consuming traditional food, particularly seal oil, is a necessary part of expressing local culinary expertise and ritual. The highest-quality seal oil is kept in the freezer and dissolves into an almost perfectly clear pool of viscous liquid when spooned on a plate. Seal oil is to Shishmaref what the truffle is to northern Italy: delicious, expensive, painstakingly produced, and dependent on a multitude of factors to develop the richest flavors and most subtle complexities. Seal oil is an indelible delicacy, and Shishmaref is both the epicenter of the art and the home of its greatest connoisseurs. Long tenured and handcrafted, food harvesting and preparation in the village has been perfected, and the resulting dishes are incredibly delicious. The carvings, the Shishmaref star, the subsistence harvests, the traditional foods, the extraordinary friendliness to strangers, and the small idiosyncrasies that are difficult to put into words make Shishmaref unique on the Seward Peninsula and unique in the world.

Perhaps unsurprising to anyone who stops to consider his or her own community, the *Kigiqtaamiut* people and life in the village are highly varied. There is great diversity among residents of the village, and while nearly everyone on the island is proudly Iñupiat, there are widely differing opinions about politics, religion, education, the role of family, the role of the government, and everything else. What all Shishmaref residents I interviewed are in agreement about, however, is that they do not want to see *Kigiqtaamiut* territory abandoned. If relocation must occur, it must occur within traditional territory.

In Shishmaref today, increased windiness and storminess, increased erosion, and diminished sea ice all threaten the low-lying island with habitual flooding. Fall storms that come off of the Chukchi Sea are particularly dangerous. These storms are accompanied by strong winds and high seas that cause erosion and bring high water. As significant ocean-facing bluffs continue to erode, the possibility of a life-threatening disaster that renders the island uninhabitable or causes a threat to life and infrastructure increases. As flooding events increase, Shishmaref residents face two distinct possibilities: they must either successfully petition government agencies or private donors to fund the rebuilding of essential infrastructure—including an airstrip, a barge landing, and a school—on nearby, tribally owned land on the mainland and along the coast; or they will eventually be forced into diaspora, scattering away from traditional homelands before, during, or after a major disaster.

A third option of relocating to nearby, tribally owned land without government aid or intervention is unlikely for two reasons.

First, it is prohibitively expensive for the small population of residents to fund the cost of building infrastructure in the US Arctic. Rural tax bases often face funding challenges for expensive infrastructure projects such as schools and post offices. Second, migration to an area without basic infrastructure—a retreat to precolonial infrastructure and economy—is unlikely because of dependence on electricity, gas, motorized vehicles, schools, medical clinics, and other nonlocal products that mark contemporary life and have since the colonization of western Alaska. High modernity expressed through education, internet use, oil and gas, and technologically advanced tools and equipment has been folded into the traditional Iñupiat lifestyle and culture in Shishmaref quite successfully in most ways. The *Kigiqtaamiut* way of life has been amazingly resilient in the face of change, particularly considering the historical tragedies of racism, colonialism, and structural violence that mark all Native American histories in this country. Today relocation options for residents are limited, and the village is vulnerable to ecological shift; but this vulnerability exists in part because of the barrage of burdens put on colonized communities.

The challenges of the colonial legacy in North America are sometimes referred to as challenges of the "fourth world"—a term used to describe Native American experiences of being colonized and "displaced" even while remaining in the same place. In Shishmaref, these challenges have included the introduction of disease, boarding schools, alcoholism, overt and subtle forms of racism, political marginalization, changes in land and resource distribution and ownership, and the increasing inability to lead a semi-nomadic lifestyle. Today, added to this long and painful list of challenges to the "fourth world" are the outcomes and effects of climate change.

What we know is that anthropogenic climate change is occurring and that the effects of warming across short, medium, and distant time scales warrant political, personal, and collective concern and intervention. The Arctic, as a land of ice, and the Iñupiat, as a people who rely on that ice, are legitimately threatened by the warming projected in recent models.[30] This is one clear reason why Iñupiat leaders have been at the forefront of the climate change debate[31] and why Shishmaref leaders and hunters in particular have a lot to teach us about climate change and its outcomes.

Rapid biophysical changes in the Arctic are substantial, and the Shishmaref community may very well be a first victim of climate change—but the village has been adapting to and surviving ecological shifts for hundreds of years. It is these ecological challenges, compounded by sociopolitical burdens, that explain why vulnerability to climate change seems so overwhelming in Shishmaref today and why community activists and residents there must work so diligently and against great odds to protect their community and determine their own fate.

This book will argue that Shishmaref actually *is* the canary in the coal mine for climate change—but not only because climate change itself is overwhelming. Shishmaref demonstrates how the negative repercussions of climate change are predicated on the gross inequity present in the world today and constructed historically across time. This story looks at climate change within the history of Shishmaref writ large and explores the grace of the *Kigiqtaamiut*, who have continuously negotiated the dynamic social landscape of indigenous people in America.

Chapter Two

Unnatural Natural Disasters

WHAT IS A NATURAL DISASTER?

When the Loma Prieta earthquake hit the San Francisco Bay area on October 17, 1989, it registered a moment magnitude of 6.9. The epicenter of the quake was near Loma Prieta peak, approximately ten miles from Santa Cruz and sixty miles south-southeast of San Francisco. It was the largest earthquake to hit the San Francisco area since 1906, and it caused sixty-three fatalities and upwards of $6 billion worth of damage. Thousands of people suffered from injuries. For most Californians living in the region, this was a "big one."

However, it appears things could have been worse. A larger, 7.9 magnitude earthquake hit Izmit, Turkey, in 1999 and caused an enormous seventeen thousand fatalities. In 2010 a 7.0 magnitude earthquake hit outside of Port-au-Prince, Haiti, and, in what was one of the most heart-wrenching challenges of the early twenty-first century, caused over three hundred thousand fatalities and continues, even five years later, to be the cause of homelessness and food insecurity. The Loma Prieta earthquake in California was powerful, and yet the fatalities associated with it were in the tens, while the only slightly larger earthquake disaster that hit Haiti caused an unbelievable two hundred twenty thousand immediate fatalities and over eighty thousand subsequent deaths associated with unsafe conditions. How can we think about these overwhelming differences?

The differences among the earthquakes of Port-au-Prince, Izmit, and the San Francisco Bay area are numerous, and they include the duration of the earthquakes and proximity to urban areas. However, the greater differences among these earthquakes were the building requirements; building types; access to quality medical care;

levels of poverty and affluence; and access to resources before, during, and immediately following the earthquake events in the different locations.

This example is important to consider because it demonstrates to us that the size of the earthquake or intensity of the natural disaster *does not necessarily predict the outcomes* of the disaster. In all of these cases the earthquake was "big," yet the outcomes were vastly different depending on the society impacted by the "big" earthquake. In these cases, the conditions that individuals, communities, and cities were in prior to the natural disaster were greater predictors of fatalities and damage than the size of the earthquake itself. We often think, almost unconsciously, of vulnerability to disasters as being exclusively linked to exposure. We think, for example, that Louisiana is vulnerable to hurricanes because it is along the coast and that Hurricane Katrina's devastation was a product of the size of the storm (possibly linked to climate change). This is an overwhelmingly common attitude, even among disaster specialists. Take, for example, the quote from Michael Brown, the Federal Emergency Management Agency (FEMA) Director, at the beginning of this chapter. "I must say," he states, "this storm is much bigger than anyone expected." The trouble with Katrina, as Brown identifies it, is the size of the storm, not the preexisting social and economic conditions of New Orleans that stem from historical racism and neglect. Social scientists, on the other hand, attribute the outcomes of Katrina overwhelmingly to the socioeconomic conditions, which preceded the storm itself.

What is important to note here is that natural disasters can very much seem like Acts of God—aberrations from normal conditions, in which an extreme natural event overwhelms society. In this mind-set, luck, prayer, preparedness, or quick thinking become the things that save some people instead of others. Even the term "natural disaster" underscores that there is no human fault at play. It is nature, not society, that causes the devastation, and nature can seem beyond anyone's control.

In reality, while no one suggests that exposure is insignificant, it just isn't the case that disaster events affect everyone equally or that skill or luck is a determining factor of survival. Quite to the contrary, when I was studying disasters in 2009 at Oxford Brooks University in England—more than a year before the devastating earthquake outside of Port-au-Prince—a professor of mine told the class that Haiti was due for a major disaster. Dr. Mo Hamza's prescient supposition did not even depend on the type of disaster—flood, hurricane, erosion, or earthquake—but on the fact that historical trauma, political marginalization, inadequate resources, land degradation, and inappropriate development created a haystack waiting for a match. Any spark would have lit the fire.

SOCIETY AND ECOLOGY: HOW HIGH WATER BECOMES A FLOOD

To better understand the nature of disasters and whom they most affect, it is first helpful to consider the diverse relationships that exist between humans and the ecologies they inhabit. These differences can make flooding, for example, a disaster for one community and a necessity for another. Flooding itself is simply a condition of high water, and conditions of high water alone do not necessarily produce negative consequences to humans. Indeed, livelihoods can be predicated on flooding. Along the Nile River, farming for five millennia was dependent on the annual overflow of the river into the croplands. While the annual rise of the Nile was controlled to some extent by earthen banks and sluices, the rise of the river into the flood plain—conditions of high water—was an integral part of the relationship between humans and their ecosystem. In similar circumstances across Northern Nigeria today, small-scale farmers exploit rich flood plains. These wetland ecosystems that are adjacent to rivers have seasonal flooding, which creates extremely fertile soils through the deposition of mineral compounds found in rivers. The rich soils associated with annual flooding have been exploited for millennia and continue to be a key resource in creating a food safety net today.[34] High water in the Hadejia-Jama'are floodplain in Northern Nigeria does not create a flooding disaster. So when does high water become a flooding disaster, like it is in Shishmaref?

Shishmaref and Northern Nigeria are each distinct ecosystems, but the ecological conditions of high water in Shishmaref produce state and federally declared disasters and in Northern Nigeria produce the necessary conditions for subsistence farming. The differences between these two situations rest in the social and economic relationships between people and ecosystems, not in the ecological conditions themselves. In Nigeria, the floodplain is farmland. In Shishmaref, current flood-prone areas are residential and commercial. Understanding disaster conditions in Shishmaref and making valid comparisons with other flood-prone areas of the world requires an intellectual unbinding of the idea of disaster from the ecological condition of high water.

This intellectual unbinding, on the ground, is actually very difficult. During interviews, I asked *Kigiqtaamiut* people to discuss the flooding events in Shishmaref. Transcripts indicate that residents responded emotionally and physically to rising water and falling shorelines—to ecological conditions, not social ones. Experiences of these flooding events are experiences of the physicality of landscape. When I asked about flooding, interviewees gave descriptions of how the land behaves and how people respond to the land as it changes during a flooding event. In this case a flood is the experience of changing ecological conditions. As one interviewee noted, "Water was just breaking off the high beach, I mean over the cliffs. . . . It was coming from both the ocean and the lagoon."[35] Floods in these interviews (and elsewhere) are frightening and unusual. "That was pretty scary," one interviewee told me, "thinking how we're going to get out of here, you know, and is it really going to flood all the way over? . . . We were on the edge of the beach watching the waves and making sure nothing was going in. I think that's when those houses were falling in. . . . We didn't think it would go that quick."[36]

These representative comments demonstrate that, in real time, disasters appear to be completely contained by the ecological conditions in the present. "The water was just breaking off the high cliffs." "We were on the edge of the beach watching waves . . . we were out there watching." The water, waves, and the breaking of high cliffs in these excerpts are the catalysts for danger.

In a disaster situation, human beings are intimately connected with changing land-scapes and are imminently threatened by those changing landscapes. The earth moves in a quake, the waters rise in a flood, the wind blows in a hurricane, and in Shishmaref the water breaks off the high cliffs—the experiences of disaster are inextricable from feeling abnormal, dangerous ecological conditions that literally move people. When asked about relocating the community in response to flooding, one respondent said flatly, "No one wants to, but Mother Nature seems like she's moving us."[37]

The experiential impression of disasters as episodic events that change the rela-tionship between people and changing ecosystems shouldn't be taken lightly. These are emotional moments. When Hurricane Rita hit my hometown of Lake Charles, Louisiana, in 2005, a month after Hurricane Katrina hit New Orleans, I went home in the immediate aftermath to help my family with recovery and cleanup. For the first week I was at home, whenever my family or I saw anyone we knew at a store, in the street, or at church, more often than not we would just cry. Helplessness is overwhelm-ing, and resorting to Acts of God or Mother Nature as an explanation for what has occurred is, in many cultural contexts, a reasonable and logical way to make sense of the overwhelming sense of loss and the upending of the status quo.

Beginning in the 1970s, however, social science researchers began to reexamine the assumption that disasters were mostly ecological exceptions to normal conditions. What they found was what I opened with at the beginning of this chapter: that social, economic, and historical conditions greatly affected how local communities experi-ence disasters. To this day, disaster researchers try to understand the relationships and interactions between social conditions of vulnerability and the ecological conditions themselves. This is the interaction between hazards and vulnerability.

HAZARD-CENTRIC VERSUS VULNERABILITY

I define hazards here as they were defined by Susanna Hoffman and Anthony Oliver-Smith in 2002, as the "forces, conditions or technolog[ies] that have the potential for social, infrastructural, or environmental damage."[38] A hazard is therefore the tornado, the drought, the nuclear power plant explosion, or the flood that causes problems for a given population. When disaster research or disaster policy is focused on the disaster itself as the singular area of study or the singular point of intervention, then that research or policy is thought of as being "hazard-centric."

Disaster prevention within a hazard-centric ideology identifies disaster solu-tions in warning systems, forecasting and prediction, and protecting populations from hazardous ecological conditions through the manipulation of ecological fea-tures. For example, in 1976 John Butler recommended the following ten strategies for flood prevention and risk mitigation: (1) forecasting, (2) levee systems, (3) large dams on rivers, (4) small dams on urban creeks, (5) river channel improvement and straightening, (6) drainage works, (7) floodways, (8) soil conservation and small dams on upper catchments, (9) flood-proofing buildings, and (10) zoning of flood plains.[39]

All but two of these suggestions (flood-proofing and zoning) focus on containing and manipulating rivers and shorelines or warning people so they can respond when a disaster approaches. Butler does not seriously consider changing human behavior or social conditions as a method of flood-proofing; things like restricting certain types of development in floodplains or ensuring evacuation processes are accessible and affordable across socioeconomic sectors.

Hazard-centric ideologies retain their currency in many federal and state disaster agencies and among the public. In almost every disaster event in America, from Hurricane Sandy to tornadoes in Oklahoma, the rally cry of "we will rebuild" and FEMA's support of rebuilding in place exemplifies the hazard-centric idea that disasters are one-off aberrations of normal conditions and that increased warning infrastructure, response plans, and technological interventions can prevent the next disaster. Rebuilding in the same way, in the same place leaves no space for reconsidering our relationship with the environment. This is true even when there is overwhelming evidence that a given location is abnormally disaster prone. Moore, a suburb of Oklahoma City, has been hit by three violent tornadoes in fifteen years. This doesn't necessarily mean Moore residents should abandon their homes, but it does demonstrate the regularity of certain "natural" hazards rather than their unpredictability and irregularity.

Like other disasters, such as the examples of earthquakes at the beginning of this chapter, tornado fatalities don't affect everyone equally. Fatalities in tornado events are 22.6 times more likely to occur in mobile home residences than in other types of residences,[40] yet only the state of Minnesota has legislation in place that requires mobile home park residents to have access to a storm shelter. Therefore, not only do mobile home residents have an increased chance of losing property and lives, but there is often no accessible method of protection for these families from death and injury that a tornado may cause. This lack of protection for people in vulnerable mobile home parks suggests an inequitable distribution of the risks across socioeconomic sectors in tornado events. In states like Oklahoma, when we know where disasters are likely to strike and whom they will likely kill or harm, it becomes surprising that intervention often aims at things like expensive warning systems and not in creating safe spaces for those residents known to be particularly at risk. Creating and legislating requirements for permanent storm shelters in all mobile home parks would change normal, everyday infrastructural conditions on the ground and would alter the behavior of residents instead of the behavior of the tornado. The fact that we aren't doing so begs the following question: Are mobile home residents in Oklahoma vulnerable to tornadoes or to poor policies and a lack of political power to demand storm shelter access?

Despite their intuitive sense and continued public appeal, hazard-centric research efforts and techno-engineered solutions to disaster began to fall out of favor among social scientists by the late 1970s and early 1980s as new research demonstrated that disasters were highly dependent on social systems and socioecological interactions.[41] High water did not always produce a disaster. A high poverty rate and inappropriate development in association with high water was much more likely to produce a disaster. For a social scientist to say that Shishmaref residents are at risk of high water or flooding is somewhat accurate—risk is certainly associated with flooding, but it is an incomplete characterization. This is akin to saying mobile home residents in Oklahoma are vulnerable to tornadoes when we know that this vulnerability could be drastically reduced with access to a storm shelter.

In Shishmaref, residents are vulnerable to a suite of consequences when high water occurs, including death, diaspora, infrastructure damage, and sociocultural disarticulation through forced resettlement. Why this community is vulnerable to those specific outcomes is complicated, but tracing vulnerability through socioeconomic processes begins to make high water seems relatively innocuous.

WHAT IS VULNERABILITY?

The term "vulnerability" is used in many academic fields as well as in common speech. Its ubiquity makes it a difficult term to pin down—so much so that Hans-Martin Füssel quotes Tinnerman as saying "vulnerability is a term of such broad use as to be almost useless for careful description at the present, except as a rhetorical indicator of areas of greatest concern."[42] *Merriam-Webster* defines "vulnerable" as "capable of being physically or emotionally wounded" and "open to attack or damage," which could certainly be applied in Shishmaref under flooding conditions but could be applied almost anywhere.

The term "vulnerability" was linked to disaster theory first in the field of geography and quickly migrated to the interdisciplinary disaster literature. There are multiple review articles about the myriad ways vulnerability is defined and its perceived usefulness.[43] Most generally, beginning in the 1980s, vulnerability has been conceived of as the conditions present in a community that include both exposure to a hazard and the inability to cope with or adapt to those hazards in a way that prevents negative outcomes, including death, infrastructure damage, and social dysfunction.

This concept of vulnerability has been used in the fields of ecology, anthropology, engineering, and economics, among many others. It is used in policy and governance forums and has become a central issue in the IPCC. The term "vulnerability" (and, following, how scientists frame an analysis of climate change) is increasingly common in any field dealing with socioecological systems in general, and anthropogenic climate change in particular. Unlike the hazard-centric approach, vulnerability approaches consider what social conditions and what socioecological relationships make hazards particularly problematic for some communities and less so for others.[44]

In figure 2.1, four conceptualizations of vulnerability are represented: (1) as a lack of entitlements; (2) as a product of political ecology; (3) as a function of pressure and release; and (4) as an outcome of exposure. Derived from Neil Adger's assessment of the existing literature,[45] these four manners of interpreting vulnerability all consider disasters to be the product of both the environment (x-axis) and social relationships (y-axis) but differ in the degree to which either is considered primary.

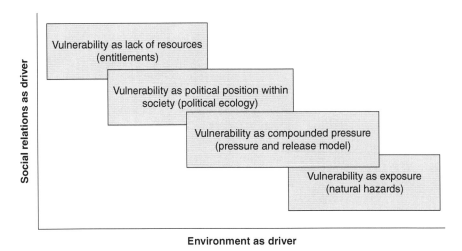

FIGURE 2.1 Conceptual diagram of vulnerability (Adapted from Adger 2006)

When the concept of vulnerability is utilized in the hazard-centric literature, it is still equated with exposure (*natural hazards* in figure 2.1). In a natural hazards framework, if you live where it is likely to flood, you are vulnerable to flooding. Flooding, in this framing, is the center of research; human interactions and social institutions are not considered the mechanism for disaster.

At the other end of the spectrum, vulnerability is associated with impoverishment, a lack of social capital, and the inequitable distribution of resources (*entitlements* in figure 2.1). Adger calls this the "entitlements" approach to vulnerability[46]—where an individual's or a community's resources and assets (social and economic) are inadequate to respond to changes or stress.

Vulnerability as a lack of entitlements can occur in the context of a natural hazard or not and is almost exclusively considered an effect of social conditions, not environmental conditions, such as is the case with death and malnutrition linked to food insecurity. In many circumstances, even when an extreme drought occurs, famine happens under conditions of lean agricultural production, not under conditions of an absolute absence of food. What this means is that even if famine co-occurs with drought conditions, yet there is still sufficient local food available to feed everyone, then famine is a product of distribution, not of the drought conditions themselves. As Adger states, "the advantage of the entitlements approach to famine is that it can be used to explain situations where populations have been vulnerable to famine even when there are no absolute shortages of food."[47] In cases where poor subsistence farmers cannot afford to buy the food that is available or cannot depend on social networks or formal institutions to provide food, disaster is a product of the lack of entitlements—not necessarily on poor agricultural conditions or an absolute lack of food. Famine can indeed occur in the absence of a drought altogether if food prices rise or seeds are too expensive to purchase. If, in any given place, agricultural and ecological conditions have to be perfect for food shortages and famine to be avoided, then these communities are highly vulnerable to famine, and any change in ecological conditions can produce famine and food insecurity *because of the social conditions present*. A researcher employing the entitlements approach would look at what social conditions produce extreme vulnerability and be much less concerned with the ecological conditions that may increase rainfall by an inch or two from one year to another.

In the Shishmaref case, under the hazard-centric framework, residents are vulnerable to death and hardship because they live on an island that is experiencing rapid erosion linked to climate change, which has, in turn, led to flooding without adequate technological protection from that flooding. Under the entitlements framework, one might consider whether or not the Shishmaref community has adequate resources (social, political, and economic) to function in a way that allows for community members to harness political will and the economic resources to fund adaptation. Vulnerability in Shishmaref under the entitlements approach is tied to resource distribution—other communities in Alaska and in the United States may be able to buy land, build seawalls, and foster political attention for safety and growth, but the Shishmaref community cannot. In the hazards approach, all of the emphasis is on the flood. In the entitlement approach, ecological conditions are merely backdrop for social relationships and the distribution of resources.

Falling between the two models discussed above, and most prevalent in the anthropological literature on disaster, is the political ecological model of disaster[48] (*political ecology* in figure 2.1). The political ecological model incorporates exposure to risk and exposure to hazards more than the entitlements approach, but it understands exposure as a function of political ecological conditions. In this model, marginalized and impoverished communities tend to live in riskier areas and have lower adaptive capacity as a result of (1) marginalization from political protection and decision-making; (2) inadequate infrastructure to cope with hazardous conditions; and (3) inadequate resources to cope with disasters before, during, and after the episodic hazard event occurs.

Susan Cutter uses the political ecological model to explain Hurricane Katrina in New Orleans. Hurricane Katrina caused significant loss of life and property and social and cultural disarticulation in and around New Orleans, Louisiana, after making landfall on August 29, 2005. Cutter explains the outcomes of Hurricane Katrina—which primarily caused harm to poor African Americans living in the city of New Orleans— as a result of development that ignored ecological conditions (i.e., the city elevation is below sea level), followed by white flight that lowered the tax base, followed by levee corrosion and disrepair linked to the lowered tax base, followed by levee failure. This infrastructural failure coincided with inadequate city planning for evacuation and the lack of individual, family, and neighborhood resources to facilitate evacuation.[49]

In this analysis, the most vulnerable communities experienced marginalization and impoverishment in multiple ways and across extended time scales, which culminated in severe negative outcomes (a disaster) when these communities were exposed to naturally occurring hazardous conditions (a hurricane). This analysis explains that certain people are vulnerable to flooding because of the inability to cope with hazardous conditions in the present, such as a lack of resources to safely and quickly evacuate. This analysis also explains why certain populations live in risky areas: a history of political neglect, white flight, and infrastructure neglect as the tax base decreased.

This political ecological model of disasters—in which a vulnerable community is recognized through a conglomeration of variables that make them vulnerable *and* located in areas of greater exposure—is instructive in the Shishmaref case study. Under this framework, we are encouraged to ask two important questions. The first is similar to that posed by the entitlements framework of disaster, namely, why does Shishmaref lack adaptive capacity or the resources (entitlements) necessary to cope with flooding and erosion in the present? The second question is, why do Shishmaref residents live in a risky location, which is repetitively exposed to hazards, in the first place?

The last model of vulnerability (*pressure and release model* in figure 2.1) is closely tied to climate change research and is adopted from the field of ecology. This model seeks to bring the environment-as-driver back into the analysis more directly—to balance the natural hazards model of disaster with the political ecological model. The pressure and release model of vulnerability understands any stress as a pressure to the system: The more pressure put on the system, the more likely the system will collapse or be forced to change into something new. Risk (of disaster) in this model is an expression of vulnerability and hazard—articulated conceptually as R (of D) = V x H. Ben Wisner and colleagues write:

> Expressed schematically, our view is that the risk faced by people must be seen as a cross-cutting combination of vulnerability and hazard. Disasters are a result of the interaction of both; there cannot be a disaster if there are hazards but vulnerability is (theoretically) nil, or if there is a vulnerable population but no hazard event.[50]

The pressure and release model is increasingly popular, particularly in climate change research and among climate change researchers. Ideally, the benefit of this model is that it incorporates root causes of vulnerability that are internal to a community and then understands hazardous conditions as an additional pressure on that community while neither underscoring nor dismissing the physical reality and importance of the hazard itself. In Shishmaref, we could say that this model would acknowledge internal conditions of vulnerability (e.g., lack of entitlements, lack of political will to demand government intervention) and the physical outcomes of flooding (e.g., breaches of seawalls, changing atmospheric conditions that lead to larger storms) as a combined explanation for negative consequences in Shishmaref when storms hit the island today. Both aspects must be systematically explored in order to understand why a disaster is imminent.

There are some important critiques of the pressure and release model. First, while the pressure and release model is successful in synthesizing social and physical vulnerabilities, it fails to provide a systematic explanation for how vulnerabilities are constructed in the first place, because it does not adequately take into account how vulnerability and risk exposure develop over time. Cutter argues that failing to understand why some people live in proximity to hazards and others do not confounds "issues within the broader context of sustainability." In other words, an understanding of sustainability is predicated on understanding the processes and events that put people in harm's way. The pressure and release model, therefore, fails to describe the mechanisms for creating vulnerability over time—even if it accurately describes vulnerabilities in the present.

The pressure and release model of disasters has become an important analytic tool for climate change researchers because hazards themselves are changing as a result of anthropogenic warming. Climate change researchers ask what happens when the environment changes so rapidly as to overwhelm previously developed social mechanisms of adaptation. Whether or not vulnerability and risk exposure are a function of history in Shishmaref is an empirical question—one that has become particularly interesting because disaster in Shishmaref is linked to climate change and because Shishmaref is used as a case study of the first "victims" of climate change. The environmental-refugee narrative suggests that Shishmaref residents are vulnerable because climatic conditions are so overwhelming and so overwhelmingly new that they are toppling social systems.

We know beyond any doubt that disasters disproportionately affect the impoverished and marginalized.[51] However, as hazards linked to climate change become increasingly unpredictable, will they overwhelm robust social systems as well? In other words, are there situations in which the hazard itself *does* exert similar outcomes on people regardless of their entitlements or political and economic positions within society? Are there situations in which new exposure to risk is so devastating that community vulnerability and political economy does not predict who lives in "risky" areas, and even the wealthy and well-connected are unprepared? Is vulnerability in Shishmaref a product of history, social relationships, and colonialism, or is vulnerability a product of an overwhelming ecological shift to which the socioeconomic, political-economic, cultural, and racial demographics of the community are circumstantial, not central?

All four models of vulnerability described above understand that the physical environment and social systems act in tandem as mechanisms for disaster. Disasters are produced when a hazard meets with a vulnerable population and produces negative outcomes and social dysfunction. Determining the extent to which vulnerability is best described in Shishmaref as a product of history, political economy, or climate change and increased exposure to hazards is a goal of this research. From here on, I employ insights from these four models to examine the case of Shishmaref, and I define vulnerability as the cumulative social and ecological conditions that put a population at risk of disaster.

STIGMATIZATION AND THE WORD "VULNERABILITY"

There is an uncomfortable irony in identifying Shishmaref residents as particularly vulnerable to climate change. I continually reference Shishmaref residents as being vulnerable to climate change, and yet whenever I do I wince internally and think to myself about Clifford Weyiouanna. Clifford often offers breakfast to people who visit or live in Shishmaref, and I took him up on it frequently because he is great company and because he makes the best sourdough pancakes I have ever eaten in my life. I dream of Clifford's sourdough pancakes.

Clifford is a man who does have a driver's license, but who definitely does not have a pilot's license, and yet somehow managed to own and learn to fly an airplane. He is, without a doubt, one of the most resourceful people I have ever met. So much so that, despite his lack of official training or documentation, stories are told about him being unofficially asked by search and rescue to fly out in his plane to find people who were missing. He would do it, too.

Clifford is unbelievably competent in his landscape. He can fix anything. He is an avid hunter and fisherman, and I have no doubt that he could quite easily survive for many years in the Arctic landscape even without modern development or technology. So, it's not without irony that I discuss Clifford, as a Shishmaref resident, being vulnerable to climate change in his environment whereas I, who have no real understanding of any basic ecosystem or resource native to the central Oregon high desert where I live, am not. How in the world is this the case?

Vulnerability is often presented as a characteristic of a community that has reached a limit, walking the razor's edge. The term is used to describe the multiple, compounding hardships that some communities in the world face, such as inadequate health care, inadequate housing, inadequate food supplies, and inadequate political representation, all at the same time. For policy workers, this image of a community on the brink is important because the response to hazards should not be one-dimensional (e.g., preventing erosion); one-dimensional responses often fail to address root causes, as Cutter explains in her analysis of Hurricane Katrina discussed above. Addressing root causes of vulnerability is the most successful way to mitigate risk in the present and the future.

However, describing vulnerability as an inherent characteristic of a community is dangerous. Describing certain communities as being inherently vulnerable can incorrectly confuse complex social relationships and environmental factors that *result* in conditions of vulnerability with an internal, trait-like characteristic that is inherent to a group of people themselves. Labeling certain groups as "vulnerable" can be stigmatizing and can result in the re-creation of outdated and racist stereotypes of indigenous peoples needing the help of white outsiders. The label can imply a lack of agency and competence. My experience in Shishmaref has overwhelmingly shown the opposite: I constantly witness competent, flexible, and resourceful individuals. The community of Shishmaref may be pushed to its limit, but the skills the community demonstrates for resilience under those circumstances are truly remarkable.

It's important to understand theories of vulnerability because vulnerability research, outlined above, is the best model for explaining this case study and the best model for understanding outcomes of climate change. Vulnerability scholars have, in the last thirty years, successfully shifted disaster conversations to root causes and, in the context of climate change, have been vocal about the inequitable distribution of burdens associated with climate change outcomes and how issues of social and environmental justice cut across the changing environment. The results of this study fall directly into this research tradition. However, the label "vulnerable" does not describe Shishmaref residents as in any way incompetent, lacking agency, or lacking flexibility and inherent resilience. Rather, it describes national and international flows of power, resources, policy, and politics that intersect the Shishmaref community and that I will describe in detail in later chapters. More insidiously, vulnerable conditions often result from the resources, land, and power that have been stolen away from certain communities in order to build resilience and wealth in other communities. If the vulnerability literature was not fundamentally applicable to this research, or as robust as it has historically been, then I would use a different analytic term.

For now, let us accept that the *Kigiqtaamiut* are not intrinsically vulnerable. They are vulnerable to a limited set of negative circumstances and events associated with flooding that are the result of complex social and ecological circumstances that implicate far-reaching global and national policies and ideologies.

Chapter Three

Flooding and Erosion in Shishmaref: The Anatomy of a Climate Change Disaster

> *The Arctic is extremely vulnerable to observed and projected climate change and its impacts. The Arctic is now experiencing some of the most rapid and severe climate change on earth. Over the next 100 years, climate change is expected to accelerate, contributing to major physical, ecological, social and economic changes, many of which have already begun.*
> —*Arctic Climate Impact Assessment*, Executive Summary[52]

CLIMATE CHANGE AND NATURAL DISASTERS

What is the link between climate change, vulnerability, and disasters? The answer is that there are many. We can imagine that as the climate shifts worldwide, places that are ideal for human habitation will also shift, meaning that some of the places we live in will become less suited to human habitation and other places may become more so. As some places become less suitable to human habitation, people will experience that ecological push through slow-onset disasters or rapid-onset disasters, the latter also known as "natural" disasters.

As temperatures slowly increase, slow-onset disasters will occur; these are incremental changes in landscapes that may at first appear innocuous but will eventually make a village, town, or city uninhabitable without significant intervention. Erosion, sea level rise, and slow-onset drought are examples of slow-onset disasters. Sea-level rise, for example, may happen in imperceptibly slow increments, but the IPCC reports that a global temperature increase of 3.2–6.2 degrees Celsius will completely deglaciate the Greenlandic Ice Sheet, causing a five-meter change in sea levels across the globe[53] and dramatically changing the coasts of the world. Roughly one in ten people alive today live in this low-elevation coastal zone.[54] To these populations, sea level rise and the increased flooding that will accompany storms means that climate change may be experienced as a natural disaster and an increase in the number of natural disasters experienced over time.

Similarly, in Shishmaref, erosion problems may not have immediately appeared to be a significant threat, but as the shoreline erodes, flooding becomes more likely. Over time, and as erosion affects higher-elevation buffer zones on the island, such as the protective bluffs that face the Chukchi Sea, flooding becomes habitual, exacerbating erosion and cutting into the amount of land on the island. These processes decrease the amount of habitable land and increase the threat of flooding in an iterative cycle that persists and increases over time.

Slow-onset disasters are not the only type of disasters climate change may bring. While incremental changes in temperature and landscape are likely, just as likely is an increase in extreme events. Rapid-onset disasters occur when an environmental limit is reached or a series of events culminate in a singular disastrous event. For example, continually warmer temperatures, longer periods of drought, and the growing pine bark beetle infestation in the American West may culminate in extreme fire seasons. Increased temperatures over the ocean, along with rising sea levels and increasing erosion, may create much larger storm surges during cyclones and hurricanes than have previously been experienced in long-inhabited regions.

Humans will experience human-caused climatic changes most acutely as natural disasters. But how are we to understand these natural disasters and their effects if, as I said before, natural disasters do not affect everyone equally? Is the flooding in Shishmaref a product of climate shift or historical disenfranchisement? What caused the earthquake disaster in Haiti?

In all theoretical models of vulnerability, vulnerable communities are a product of social circumstances and ecological features in the landscape as well as the interactions between those systems. In the case of Shishmaref, Alaska, the village has been identified as a case of environmental migration linked to climate change. If ecological features and social circumstances are interlinked in disaster scenarios, then in order to understand the construction of vulnerability in Shishmaref, it is imperative to investigate the linkages between cultural, social, and ecological systems—particularly those systems that are in flux. This chapter focuses on the ecological changes that contribute to vulnerability in Shishmaref.

CLIMATE CHANGE IN SHISHMAREF, ALASKA

The Arctic is a stunning and a complex environment in which to study climate change. Polar amplification; the attendant changes in weather and ice patterns; and the implications for landscapes, animals, and people have made the Arctic a hotbed of climate change research in the last thirty years. There is now a robust literature on how atmospheric, terrestrial, and hydrological systems have changed over time, linked to both greenhouse gas emissions and natural processes.[55] Indeed, "Alaska has been called a 'climate canary' because it is already seeing the early effects of global climate change."[56]

In spite of this robust literature, climate change modeling and research remains difficult to downscale to any specific locale, in part because of modeling limitations.

While recently there have been advances in climate model downscaling,[57] it remains difficult to link large-scale environmental change and climate change research across the Arctic to a particular coast, lagoon, riverbank, or community. In this case, understanding how Arctic climate change trends affect Sarichef Island and the Shishmaref coast is not straightforward.

What is certain is that climate scientists have documented substantial changes in the Arctic climate regime over the last one hundred years, with increasing changes recorded since the 1970s. The following research demonstrates that Shishmaref residents have also observed and documented in the oral record significant ecological changes over time, and particularly within the last thirty to forty years. Tables 3.1–3.5 present ecological changes observed by Shishmaref residents at the local scale and contrast and compare these observations to scientific findings on climatic changes in the Arctic.

The observations made by Shishmaref residents are not necessarily a comprehensive set of changes observed by residents on the landscape but are those changes that came up during the interviewing process. The tables catalog changes in the climate (e.g., stronger currents on the ocean side of the island) that were identified by two or more individuals during interviews. To compare Shishmaref residents' observations with scientific data on climate change in the Arctic, these tables incorporate the framework adopted by hydrologist Larry Hinzman and his coauthors in their 2005 article,[58] which summarizes Arctic climate change research findings with a particular focus on Alaska.

There are five broad categories of change that were most salient in the observations made by Shishmaref residents and that corresponded to published scientific data on climatic changes throughout the Arctic. These categories were changes in weather, permafrost thaw, thermokarst ponds, freeze-up, and coastal erosion. These categories are interrelated, particularly in interview data from Shishmaref, so that permafrost thaw and erosion, for example, are co-occurring, mutually constituting phenomena. They are separated out here for comparative purposes.

WEATHER

Weather throughout the Arctic has been observed to be increasingly unpredictable. In my interview data, unpredictability was tied specifically to ice and wind conditions. Ice unpredictability and a decrease in ice thickness, like weather unpredictability in the Arctic climate change literature, are recognized by Shishmaref residents as creating hazardous travel conditions. Shishmaref residents particularly identified increased windiness, warmer winter temperatures, longer fall seasons, and fluctuations in winter wind direction as changes that have occurred within one lifetime. Table 3.1 summarizes overall weather changes observed in Shishmaref and compares these changes to weather changes observed by scientists across the Arctic.

TABLE 3.1   Changes in Weather Patterns

Climate change observation comparison	Shishmaref interviews	Hinzman et al. 2005
Theme: Weather	Changes in weather and ice	Weather changes
Evidence	Stronger winds, changes in winter wind direction (consistently north winds in winter now—used to be more variable); spring and fall longer, winter shorter	Greater variability, less predictable weather
Effects	Erosion along the island, sea ice changes	Increased mortality to plants and animals, greater hazards in traveling (Krupnik and Jolly 2002[59]; Simpson et al. 2002[60])
Location		North America
Climate Driver*		Changed synoptic patterns
Discrepancies	Wind and temperature were specifically referenced—weather variability may be implied, but I did not specifically ask about variability, and interviewees did not specifically identify increased variability	Stronger winds and changes in wind direction were not mentioned
Time Frame	Recent decades—within one lifetime	Recent decades

*In Shishmaref, this is expressed as co-occurring features.

PERMAFROST THAW

Thawing permafrost has been consistently observed by Shishmaref residents and Arctic climate change scientists. In Shishmaref and throughout *Kigiqtaamiut* territory, residents constantly engage with and observe the landscape. Permafrost thawing is experienced, not just observed, and changes in time are marked by personal histories. For example, Clifford Weyiouanna remembers building his reindeer corral thirty years ago, hitting ice at one foot below ground level. Today he can dig much farther without hitting ice. Fred Eningowuk had to move a cabin on Serpentine River because shifting permafrost caused infrastructure damage. Changes in permafrost have been swift. Residents report that when permafrost and ground ice are exposed to the ocean, erosion processes speed up exponentially. Permafrost thaw is also linked to draining tundra lakes. Table 3.2 outlines these observations.

TABLE 3.2   Changes in Permafrost

Climate change observation comparison	Shishmaref interviews	Hinzman et al. 2005
Theme: Permafrost	Permafrost thawing	Permafrost thawing
Evidence	Easy to dig into the ground, which used to be frozen; visual changes in landscape; cabins sinking	2–4 degrees Celsius warming; thawing (Osterkamp and Romanovsky 1999[61]; Clow and Urban 2002[62]; Romanovsky et al. 2002[63])
Effects	Exposed permafrost "ice" at the coastal shoreline that, following exposure, rapidly erodes; have had to move cabins and camps	Thermokarst, infrastructure damage
Location	Erosion noticed particularly at Cape Espenberg hills, at Serpentine, and on the ocean side of the island, but many people point out that erosion happens on both sides of the island— linked to permafrost thaw	Alaska
Climate Driver*		Warmer air temperature, changes in snow
Discrepancies	Snowfall has not changed significantly	
Time Frame	No longer than two generations	Since the late 1800s, especially last decade

*In Shishmaref, this is expressed as co-occurring features.

THERMOKARST PONDS

Table 3.3 identifies an effect of permafrost erosion and anthropogenic warming that is of particular importance to the Arctic and sub-Arctic and may have important effects on hydrological regimes on the Seward Peninsula, such as the availability of fresh water:

> The important processes involved in thermokarst include thaw, ponding, surface and subsurface drainage, surface subsistence and related erosion. These processes are capable of rapid and extensive modification of the landscape, and predicting, preventing or controlling thermokarst is a major challenge for northern development.[64]

TABLE 3.3   Changes to Thermokarst Ponds

Climate change observation comparison	Shishmaref interviews	Hinzman et al. 2005
Theme: Thermokarst ponds		
Evidence	Three big lakes that emptied out: "There was a little creek attached to them. I think the permafrost melted and drained them out."	Decrease in area
Effects	New channels, landscape changes	Landscape and vegetation changes (Yoshikawa and Hinzman 2003[65])
Location	Approx. 5 miles west of Sarichef Island	Seward Peninsula, Alaska
Climate Driver*	Permafrost thawing	Degradation of permafrost
Discrepancies		
Time Frame	Last few years	1951–2000

*In Shishmaref, this is expressed as co-occurring features.

Thermokarst is not a commonly used word in Shishmaref, but residents have observed large ponds that have completely disappeared and new channels draining into the ocean where these ponds may be draining through. This kind of extreme topographical change that is happening quickly enough for residents to observe in a single lifetime—or even within the span of a single year—corresponds with hydrological data suggesting rapid changes to the water regime on the Seward Peninsula.

FREEZE-UP

Shishmaref residents repeatedly indicate that the ocean and lagoon freeze later than they have in the past. Freeze-up and spring break-up are momentous occasions on the island because the in-between states of water on the ocean and lagoon (not completely frozen, not completely thawed) "trap" people on the island and prevent easy travel to and from the mainland or out into the ocean to look for sea mammals. This means that freeze-up and break-up dates are remembered and recorded. Clifford Weyiouanna remembers his father consistently traveling across the lagoon with a dog team on his birthday. This sets the freeze-up date of the lagoon to October 22, which can be measured against freeze-up dates today. In my interview set, Shishmaref residents did not

TABLE 3.4   Changes to Freeze-up and Break-up

Climate change observation comparison	Shishmaref interviews	Hinzman et al. 2005
Theme: Freeze-up/ break-up	Later freeze-up	Later freeze-up, earlier break-up
Evidence	Later freeze-up of lagoon and ocean ice: "It freezes in December; when I was young it started to freeze in October"	Earlier break-up, delayed freeze-up (Magnunson et al. 2000[66]; Rühland et al. 2003[67])
Effects	Thinner ocean ice—thin ice can mean dangerous conditions for hunters; difficult to travel by snow machine on the ocean or across the lagoon	Longer open-water season, changes in aquatic ecology, riverine transportation
Location	Chukchi Sea and Shishmaref Inlet	Lakes and rivers in the northern hemisphere
Climate Driver*	Warmer temperatures in winter, longer falls and springs	Warmer air temperatures
Discrepancies	Ocean and lagoon freeze-up and break-up discussed more than the rivers	Does not list ocean or lagoon ice as having later freeze-up
Time Frame	In two generations—when the lagoon was consistently frozen by the end of October	1900s to present

* In Shishmaref, this is expressed as co-occurring features.

discuss the freeze-up and break-up of river systems—though this does not suggest that the freeze-up and break-up of rivers has not changed. Hinzman et al.'s summary of the literature discusses river freeze-up and break-up exclusively, not sea ice or lagoon ice. I have combined and compared these observations in table 3.4 because they are related to similar climate drivers but are observations of different hydrological systems.

COASTAL EROSION

Finally, Shishmaref residents are experiencing coastal erosion: the rapid loss of land and shoreline as bluffs on the island and hills and bluffs along the mainland coast disappear, which is associated in the climate change literature with warming

TABLE 3.5   Changes in Coastal Erosion Patterns

Climate change observation comparison	Shishmaref interviews	Hinzman et al. 2005
Theme: Coastal Erosion	Heavy erosion rates throughout the coast	Coastal Erosion
Evidence	Increasing rates of erosion throughout *Kigiqtaamiut* and *Tapqagmiut* territory	Increased erosion rates (Osterkamp et al. 2000[68])
Effects	Seawalls become necessary, relocation	Increased sediment and carbon flux to ocean, infrastructure damage
Location	Sarichef Island, Serpentine River, Cape Espenberg	Barrow, Alaska
Climate Driver*	Stronger current, permafrost thawing, increasingly violent winds	Shift of storm winds, active submarine erosion
Discrepancies	Stronger currents mentioned in multiple interviews	Not mentioned
Time Frame	Increasing since 1974	1949–2000

* In Shishmaref, this is expressed as co-occurring features.

temperatures in the Arctic. Large-scale erosion of cliffs along the mainland coast, especially at Cape Espenberg, is also reported. Coastal erosion on Sarichef Island is linked directly to migration outcomes; as Sarichef Island diminishes, the chances of flooding increase and permanent inundation of the island with floodwaters becomes more likely. It is important to point out that coastal erosion is not confined to Sarichef Island, although erosion on the island causes the greatest risk to residents. Table 3.5 compares climate change research with local observations.

CLIMATE CHANGE IN SHISHMAREF: A DISCUSSION

There is widespread ecological change occurring in the Arctic as a result of anthropogenic climate change, and Arctic residents observe and respond to these changes. When I spoke with Shishmaref residents, one of the more overwhelming experiences for me was the level of detail and specificity with which most people spoke of the landscape and of changes in the landscape. Very rarely were statements generalized and they were even less likely to be cataclysmic or propagandistic. More often statements were supported by personal experience, exact location, and precise detail. The situated nature of observation and experience in Shishmaref about climate change and ecological shift is consistent with that of other Iñupiaq groups, and this precision compels anthropologists to take the oral record seriously, particularly when interpreting the "grounded truths" of scientific statements.[69] Local frameworks for interpreting changes in the landscape, however, are place-specific discourses that often but not always conform to the "climate change" discourse directed by scientific or policy norms, and the label "climate change" can diminish complex, grounded local knowledge.[70] It is illuminating to consider these discourses as both parallel and interacting narratives assessing the same contours of landscape, identifying similar but slightly divergent experiences and phenomena as appropriate "data" for interpretation. Though it is important to understand and acknowledge the differences, the similarities in these discourses are apparent in the tables above.

That the landscape is changing and that this is linked in part to greenhouse gas emissions and anthropogenic warming is beyond doubt or reproach, but it may be that coastal erosion is a product of the interaction between anthropogenic climate change and seawall construction. Trying to determine the extent of coastal erosion that is caused by anthropogenic climate change (instead of development or natural processes) is problematic, in part because there is no research that separates the drivers of coastal erosion and in part because the outcomes and experiences for residents are not different whether erosion is a natural process, an outcome of development, an outcome of climate change, or (as is most likely the case) a combination of these three factors. Imagine trying to parse out the drivers of change when the bluff in figure 3.1 meets the storm in figure 3.2 and threatens the house in figure 3.3 with flooding.

FIGURE 3.1   Bluff erosion in Shishmaref (Photo courtesy of Tony Weyiouanna)

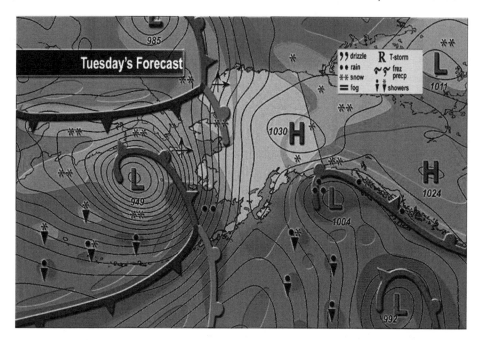

FIGURE 3.2　Weather map of 2011 "Super Storm" (Map National Weather Service "Bering Strait Superstorm," November 9, 2011)[71]

FIGURE 3.3　The end of the seawall and a house in Shishmaref (Photo by Elizabeth Marino)

Nevertheless, the following section engages a conversation about the drivers of erosion as a gateway for understanding how a combination of ecological and social processes lead to climate-change-related disasters.

WHAT IF ANTHROPOGENIC CLIMATE CHANGE IS NOT CAUSING EROSION IN SHISHMAREF?

In a provocative new chapter on Shishmaref erosion, dissenting archeologist Owen Mason makes the following claims:

> The prevailing narrative from Shishmaref represents it as "the front line" of cli-
> mate change. . . . Shishmaref *does* [original emphasis] face a duel threat, both
> from coastal erosion and from the thinning and disappearance of sea ice that may
> cripple its subsistence economy. . . . Missing from the media and community con-
> versation is that the 1 km-long bluff on which the modern village is concentrated
> is a developed coastal reach that has been subject to nearly 75 years of erosion
> control efforts and that its erosion history differs significantly from that of adja-
> cent undeveloped coasts on the Seward Peninsula. In terms of historic erosion
> processes, Shishmaref more resembles some areas of the New Jersey shore and is
> better understood as a battle in the ongoing "war" between the US Army Corps of
> Engineers (ACE) and the shore.[72]

With this assertion, Mason contradicts the prevailing theory that climate change is driving migration and erosion in Shishmaref; but seen differently, this assertion introduces the complex and sociohistorical nature of climate-change-related disasters. In Shishmaref, erosion on the island has been assessed as a product of development, which insulates and warms the ground under infrastructure;[73] increasing temperatures linked to climate change, which move the permafrost boundary north;[74] diminished sea ice, also a product of climate change; *and* inadequate and ineffective seawalls, which have increased erosion on the island.[75] It is difficult to determine where the outcomes of atmospheric temperature increases end and where the effects of development and human intervention begin.

Mason's article points out that even the extent of erosion itself is unknown along most of the Alaskan coast. In Shishmaref, scientists are inconsistent when estimating how much land has eroded over the last one hundred years. An Army Corps of Engineers report, for example, estimated the erosion total for the last thirty-one years at a rate that was 57 percent higher than a report from Colorado University.[76]

Mason points out that erosion rates are highly variable across decades and that ero-sion rates were highest in the 1970s (when relocation was first being discussed [accord-ing to Percy Nayokpuk]) and in the early 2000s (when interest in relocation began again in earnest and the community relocation vote occurred). Mason's assessment is that seventy-five years of development and intervention has increased erosion rates on the island compared to undeveloped coasts. In fact, Mason claims that intervention aimed at protecting the island, such as particular revetment and seawall projects, may have actually increased erosion rates on unprotected parts of the island by intensifying and redirecting wave action and wave energy away from developed areas of the island.

Mason's article arguably presents the best scientific data on erosion, storm action, and erosion protection for Shishmaref to date. What his article implies is that the link between climate change, erosion on the island, and environmental migration lacks sufficient substantiation within the scientific literature. Mason's review of the scientific data, including his own field notes, implies that climate change is not the driver of coastal erosion on Sarichef Island.

The Army Corps of Engineers disagrees with Mason's assessment, writing, "Climatic conditions have led to icepack development occurring later and later each year. Without the icepack in place, the island is more susceptible to fall and early winter storms that have *increased* [my emphasis] erosion and littoral drift."[77] The Corps estimates that erosion rates in Shishmaref are in the process of increasing and that it is continued erosion that threatens inundation and flooding—risks that would essentially destroy critical infrastructure in the village. The map in figure 3.4 was developed by the Corps to show projected coastline erosion under current conditions. The Corps' report suggests that climate change is an important driver of erosion in Shishmaref.

Most likely both the Mason and Corps reports are correct. Climate changes will interact with other processes to create dynamic conditions. In Shishmaref infrastructure

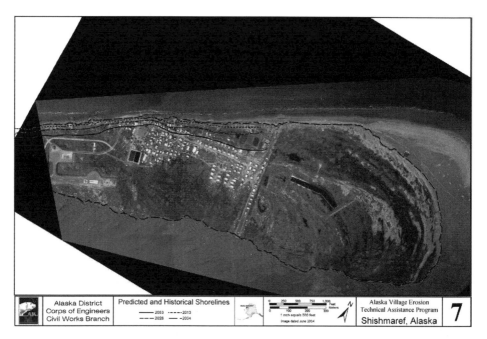

FIGURE 3.4   Map of predicted and historical shorelines of Shishmaref, Alaska (Alaska District, Corps of Engineers, Civil Works Branch)

and poor development decisions interact with changing ecological conditions. Mason points to two compelling pieces of information: first, that seawalls and other revetment projects were at times more harmful than helpful; and second, and most importantly, that coastal erosion is a natural process of barrier islands. He suggests that erosion would have happened in any climate scenario. Interview data corroborates this claim.

WE KNEW THE ISLAND WOULD DISAPPEAR, SO WHY DO WE LIVE HERE?

In multiple interviews, Shishmaref residents repeated a common local dictum that says that the Shishmaref barrier island chain is going to disappear into the ocean and that elders of the *Kigiqtaamiut* people have always known this would be the case. In an interview with one Shishmaref resident, she reported:

> My grandparents used to talk about it. Even their parents used to say, when you guys get older you're going to see big storms; you're going to see our land get smaller. And when our grandparents lived long enough to see that happening they say, "our parents told us about this." Some of them didn't even want to be buried here even on the island. [They said] when I die, will you please bury me somewhere else, not here.

Fred Eningowuk told me one day, "The elders always knew the ocean would take back this island, take back what it created."[78] Multiple individuals made similar statements in other interviews. While residents observe climatic and terrestrial changes in Shishmaref, and while coastal erosion *has increased in speed and extent* throughout *Kigiqtaamiut* territory (likely due to permafrost thaw and the effects of a warming climate), residents recognize that the Chukchi Sea Coast and specifically the Shishmaref barrier islands are a fluctuating and impermanent landscape.

If the *Kigiqtaamiut* knew that the barrier island was impermanent and would subsequently be at risk for flooding and erosion as the ocean "took back what it created," then why do people live there? In July 2008, at the beginning of this research, Tony Weyiouanna relayed the statement with which I'll open the next chapter: "People aren't talking about the past, about why villages were here in the first place. And they're not talking about the future—what it's going to be like for our kids."[79] Trusting his expertise in this early interview, I added a series of questions to my subsequent interview scripts regarding where interviewees and their ancestors were born and what made them relocate to Sarichef Island and the village of Shishmaref permanently.

Overwhelmingly, residents answered the question, "Why did you, your parents, or your grandparents move permanently to Shishmaref?" in one of three ways: (1) this is a good place to hunt sea mammals and have access to the mainland, (2) the Bureau of Indian Affairs (BIA) built a school here, or (3) this is a good place to hunt and the

BIA built a school here. Most often, my interviewees gave the third answer, such as the following:

> Well, it's always been a traditional village . . . this has always been a good central place to hunt. Our community is mainly built for seal hunting. This has always been a very good place to access the ocean during the spring, during the fall. And then from here we can travel to the river and then up and down the coast. So, people originally moved here because it's a good location. Plus, at the same time, the school was built here.[80]

Relocation and settlement linked to education was in part a decision residents made—but under constrained and mandated legal requirements. Fred Eningowuk points out, "They moved here permanently because of the school, BIA school, required everybody to go to school and so this became a permanent settlement, otherwise there were other settlements up and down the coast.[81] An interview with Tommy Obruk gives insight about the productivity of the island:

> Shishmaref was kind of spread out, long ago, before the school and the church. From Cape Espenberg to Ikpik. After the school and the church came they decide to have Shishmaref [in the] central part. Like I said, it was the elders that decided for their families, you know, where it was easier for them to hunt. North sea for springtime hunt and for fishing and seal hunting in the lagoon, and moose and salmon berries and fish nets, berry picking or mostly up in Serpentine flats and I think that's why they chose island of Shishmaref.[82]

Ecological changes linked to climate change in Shishmaref are significant and include dramatic changes in permafrost, later freeze-up and earlier break-up, changing weather patterns, and increasing erosion, which leads to flooding and pushes migration. These climatic changes, however, happen against a backdrop of naturally occurring ecological conditions (such as the dynamic coast of Alaska) and happen in conjunction with human error (such as building seawalls that ultimately create even greater rates of erosion). Mason did not debunk climate drivers—he pointed out, instead, what the vulnerability literature would predict—that Shishmaref is in harm's way because of a unique set of social and ecological circumstances happening in tandem. These circumstances are constructed over time and through colonial processes. No one knows how early-development decisions were made, but the significance of the school as a mechanism for relocating family groups that were "spread out" along the coast is a common historical theme throughout the colonial period across North America. This chapter focused on the ecological drivers of climate change and erosion. The next chapter focuses on the construction of vulnerability through time.

Chapter Four

Seal Oil Lamps and Pre-Fab Housing: A History of Colonialism in Shishmaref

People aren't talking about the past, about why villages were here in the first place. And they're not talking about the future—what it's going to be like for our kids.
—Tony Weyiouanna, interview, July 2008[83]

In Shishmaref, residents point out that permanent settlement in the village is linked to the construction of the school and legislation that mandated school-age children to attend. Western infrastructure development was explicitly used by missionaries and US government leaders to promote colonial institutions and to discourage traditional infrastructure, traditional patterns of mobility, and traditional institutions. Before the colonial period, however—before Sheldon Jackson and the post office and seawalls and climate change—first came the island.

THE ISLAND IS A CENTER OF SUBSISTENCE

The history of Shishmaref begins with the centrality of the island. Nearly everyone interviewed, including relocation activists and community organizers—people who were adamantly pro-relocation—were saddened by the idea of leaving Sarichef Island and the contemporary village of Shishmaref. At some point during interview sessions, without prompting, many interviewees made note that Shishmaref was a perfect access point for sea mammals, especially bearded seals, or *ugrut*—the subsistence foods through which, by hunting, storing, and eating, Shishmaref people express their cultural vitality most publicly. A vast majority of Shishmaref residents want to relocate the village in subsistence territory on the mainland, yet residents acknowledge that they would have to return to Shishmaref and pitch tents in the springtime to conduct the seal hunt. Moving farther away from the sea is a concern for some residents. Even when individuals advocate for relocation, some worry that life will be increasingly difficult off of the island. As Fred Eningowuk has said,

If we were to move to the mainland it's going to be a lot harder to live the way we are living right now because we subsist off the ocean, the land, the lagoon. Come the springtime if we move to the mainland we're going to have a lot harder access to the ocean to do our spring hunt. Usually that time is when the ice is, the lagoon ice is not very safe to travel on to get to the ocean. . . . I think we would have a lot of accidents with these younger generations trying to get to the ocean.[84]

Sarichef Island is located five miles from a fresh water source on the mainland. The island's proximity to the ocean and the mainland allows travel up and down the coast to hunt seals and other sea mammals (such as walrus) and allows access to river drainages and caribou hunting locations farther inland. Traveling on the mainland, residents can access land mammals, river fish, and greens and berries; although these are important, there is no doubt that Shishmaref residents are and have been oriented toward the sea. The ancestors of the *Kigiqtaamiut*—the greater political and geographical nation, the *Tapqagmiut*—were coastal people and marine mammal hunters. The following summarizes a history of their profound and resilient cultural legacy.

A HISTORY OF THE ISLAND PEOPLE

The Seward Peninsula coast has always been host to a rich, complex diversity of cultures, technologies, economies, and ideas for thousands of years prior to the whaling traditions that brought Russian Cossacks and European whalers to the northwest coast of Alaska. The Arctic Small Tool Tradition—which lasted for over three thousand years (approximately 2900 BC to 1000 AD) and is associated with diverse economic strategies and technological expertise[85]—is an example of that rich history.

The Seward Peninsula and the Bering Strait region in general are known as being the most significant migratory access point into North America—the first point of contact in North America after Asian migrants crossed the Bering Land Bridge. But instead of viewing the Seward Peninsula as a permanent migratory route, Giddings stresses that "the emphasis can be, for a time, on the *cultural stability* of a Bering Strait which is a center, rather than a way-station, of circumpolar ideas."[86] The Bering Strait as a region has been consistently inhabited, and archeological records demonstrate continuous technological advances and extensive trade routes. This allows for both cultural stability and dynamic change. Patterns of mobility also exhibit the characteristics of stability and dynamism in concert. Changes in the landscape, including unstable sea levels, fluid coastlines, and the destruction of village sites, have been reoccurring conditions,[87] and communities for thousands of years have adapted to social and ecological shift by making selective changes and maintenance to social and cultural habits, technologies, and customs. Traditional mobility patterns throughout the northwest coastal region of Alaska demonstrate the fluidity of change and tradition.

By the nineteenth century on the Seward Peninsula, Iñupiat people along the Bering Strait were sedentary seasonal. Movement was governed by seasonal employment;[88]

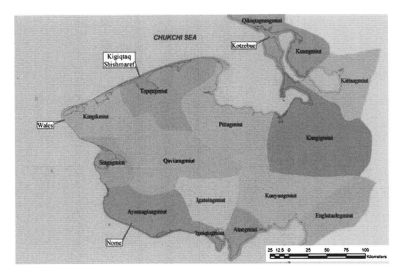

FIGURE 4.1 Map by Josh Wisniewski (2011), based on work by Burch (2006)

seasonal rounds and human migration were determined by animal movements and availability,[89] by ice conditions, and by the weather. Shishmaref residents today are mostly descended from the *Tapqagmiut*. The *Tapqagmiut* people were a loosely joined "nation"[90] of family groups,[91] who shared dialects, lands, and feasting periods and festivals throughout the year. Figure 4.1 shows the "nations" of the Seward Peninsula in the nineteenth century, with the island and Shishmaref identified in the northwest corner of the peninsula.

Nations in the Seward Peninsula stayed within their respective territories for most subsistence activities, and seasonal rounds differed between nations. Some Seward Peninsula Iñupiat nations moved inland for fall and winter. For the *Tapqagmiut*, fall and winter settlements were located along the coast. At freeze-up (or possibly earlier), smaller family groups would gather at a larger, more permanent village site and remain there through break-up.[92] People were not immobile during the winter and would travel inland for caribou hunting, but overwinter villages were more stable places to gather. Housing structures in these villages, which we will discuss later in this chapter, reflected greater permanence and engineering complexity.

During break-up, at the height of the *ugruk* hunt, *Tapqagmiut* would move out along the coast—including moving onto the shore ice itself to hunt for seals. "So all the camp sites that we had along the coast were based on what the ice conditions were going to be. But those days . . . just by looking at the ocean ice you could pretty much predict what the ice was going to do" (Clifford Weyiouanna, July 21, 2008).[93] During the spring and summer, *Tapqagmiut* families spread out over their land for inland hunting and fishing. Burch estimates that the population of the Shishmaref region in 1800 was about 510,[94] which is not a significantly different population than inhabits the island today.

Mobility throughout the year, while patterned, was dynamic. As the quote above by Clifford Weyiouanna indicates, decisions about movements, camps, and mobility were made following an analysis of weather and ice conditions. The particularities of any given minute, day, season, or year could significantly influence where a small family group or larger family unit would move to and whether or not they would gather or disperse. High mobility therefore allowed for flexibility regarding weather conditions.

Kigitaq, or "Old Shishmaref," was the largest winter settlement in the *Tapqagmiut* region and was located on Sarichef Island. Formal archeological excavation has not been carried out on the island itself; but artifacts found on the island by residents have been dated to 1400–1500 AD.[95] The map in figure 4.2 identifies the contemporary village of Shishmaref (labeled Shishmaref) and the site of "Old Shishmaref."

"Old Shishmaref" sits on a low sand bluff on the east side of the island. Amazingly, even though Old Shishmaref sits at a low elevation and is close to the water, the traditional village site is on an area of the island that has not experienced significant erosion[96]—demonstrating the deep ecological knowledge that drove *Tapqagmiut* and *Kigiqtaamiut* decision-making.

According to Susan Fair, the Iñupiaq designation "*Kigiqtaamiut*" traditionally referred to families who were identified with this overwinter village.[97] The *Kigiqtaamiut,* or "people of the island," were a subset of the *Tapqagmiut* citizenry who used the island

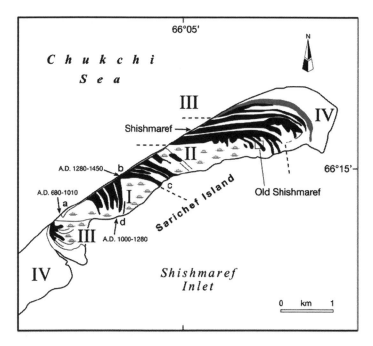

FIGURE 4.2   Sarichef Island historical development (Map of island taken from Mason et al. in press, radiocarbon dates from Mason 1996)

as a sea mammal staging ground. Other overwinter villages include *Ikpik* and areas around Cape Espenberg.

Families in Shishmaref today are still associated with the traditional village sites most utilized by their ancestors. Localized knowledge, including knowledge of place names and landscape, is linked to specific family groups and where they spent the winter prior to settling on the island. Today, traditional land tenure is loosely maintained in the village, and both formal and informal rights to hunting and gathering locations are dependent on family histories.[98] For example, families originally from the *Ikpik* area down the coast or families from up the coast at and near the Cape Espenberg area return to these places to hunt, fish, and gather. In the following excerpt Clifford Weyiouanna identifies specific people and families who have locally recognized access to and knowledge of traditionally inhabited areas. "S—, he knows all the names from Serpentine east, every little hill, every little creek. And you take the R—, and they're all on the west side—Ikpik. They know that area real well. I grew up in Arctic River—our families had special areas that they went to.[99]

In spite of continuing to identify specific families with early, precolonial (and post-colonial, as some *Ikpik* families didn't move into the village until the 1950s) settlements, the Shishmaref Relocation Coalition website identifies all Shishmaref residents today as *Kigiqtaamiut*, or people of the island. This is another example of selective social adaptation and a public display of community cohesion. While land tenure to some degree is maintained through continuous family use of traditional hunting and fishing grounds, Shishmaref people today recognize themselves under a single Iñupiaq place-name-based designation: the *Kigiqtaamiut*.

The economic history of the nineteenth century tells us that prior to colonization and sedentarization, the *Tapqagmiut* were scattered in smaller family groups throughout the region and would gather in smaller villages to overwinter. High mobility was an economic strategy, as *Tapqagmiut* people followed their resource base inland during the summer and onto the sea during sea mammal hunts. High mobility therefore allowed for flexibility. Seasonal migration was deeply connected to reading the weather, the animals, and the ice and to knowing where one should be in relation to environmental, terrestrial, and oceanic conditions. One's family group had patterned migration practices, but at any moment these patterns could change because of the weather. As such, this high mobility constituted an extremely successful adaptation strategy to flooding and erosion.

MOBILITY AND ADAPTATION

If Shishmaref has been inhabited for at least five hundred years, and if the coast has always been dynamic and impermanent, then why was it not risky to live there in the past? There are two probable answers to this question that are suggested from the literature and from my interviews. First, as long as the island has been in use, it has never been inundated by water. Second, high mobility was a successful adaptation strategy to protect against flooding and erosion[100] because movement off the island could be quick and efficient and infrastructure losses were minimal.

In interviews and in casual conversation I routinely asked whether or not there were old stories of flooding—from before the school or post office was built—and the answer was always no. This is not to say that *Tapqagmiut* people did not experience high water. When unusual flooding hazards or high-water events occurred prior to sedentarization, it appears that people simply moved "to higher ground." In these cases, conditions of high water did not translate into a flooding disaster. Now think about this: This means that a socioecological relationship existed in which high water was not a disaster, was not problematic. In multiple interviews with elders, stories of dramatic ecological events that created high water were discussed without reference to subsequent disastrous outcomes. I asked Tommy Obruk about his memory of flooding disasters, and he told me, "Twice, I think, I witnessed a tidal wave. One at fall time, when they were in skin boats and we flooded up the river. We had to move to higher ground, up in the hills."[101]

Mobility and retreat to higher ground or away from the coast seems commonplace in the Seward Peninsula. Sister Anna Huseth, a missionary from Minnesota who was stationed in the nearby village of Teller from 1919 to 1928, writes about a seasonal migration linked to high water. Again, this does not appear as a catastrophic or even inconvenient event; it rather was incorporated into the seasonal round:

> Our little village, when the spring break-up comes, is flooded so that we must move out. We pack provisions and tents and go inland to camp where we fish and hunt so as to get our winter supply of food ready.[102]

The ability to "move to higher ground" is a kind of flexibility that prevented high-water events from becoming flooding disasters. Sister Anna Huseth referenced flooding as habitual but not problematic. Flexibility through mobility in this context refers not only to the movement of people but also to the mobility of equipment, housing, cultural meaning, and social practice.

High mobility in these cases is directly linked to infrastructure. The three housing structures used along the coast all promoted high mobility and therefore promoted an astounding flexibility to dynamic coastal conditions. These distinct structures for the *Tapqagmiut* were (1) the tent, made of wooden poles and seal or other skins; (2) the icehouse, constructed while hunting on or off sea ice; and (3) the more permanent sod house.[103]

We can say, then, that dwelling structures on the Bering Strait, up until the twentieth century, constituted an adaptive strategy to the fluctuating coast. The construction materials—timber, skin, and sod—used to build critical infrastructure on the Seward Peninsula through the nineteenth century were largely available. The skill sets needed to construct dwellings were part of a local repertoire of knowledge. Local materials and in-group knowledge were the "investments" in infrastructure, and this infrastructure was transferable among residents, to different hunting grounds, and in deference to changing weather conditions, social conditions, and dynamic coastal conditions.

Social habits and seasonal rounds were thus incorporated into patterns of high mobility. Iñupiat life prior to the turn of the twentieth century was on the move—and being on the move was a socioecological relationship with the immediate environment that gave residents a significantly reduced vulnerability to flooding. Just as residents around the floodplains of the Nile incorporated high-water events, so early Iñupiat lifestyles were flexible enough to respond to high water without creating a disaster.

SCHOOLS, HOUSES, AND SEAWALLS: THE CRITICAL DEVELOPMENT OF COLONIALISM

> *The old heathen home, from its very character, was a hot-bed of vice.*
> —Northern Canadian Methodist Missionary Thomas Crosby, 1907
> (quoted in Perry 2003)[104]

The first written account of *Tapqagmiut* people engaging with non-Native explorers was from July 4, 1816, when Otto von Kotzebue landed on Sarichef Island and observed semi subterranean houses (presumably in *Kigitaq*) and named both the island and Shishmaref Inlet.[105] As the sailors came ashore, the *Kigiqtaamiut* present in the village retreated, though some members of the expedition later met with hostile Iñupiat in the same area[106] who launched projectiles toward the Russian sailors.[107]

For the next two hundred years, and particularly in the last one hundred years, *Kigiqtaamiut* and *Tapqagmiut* social life would change profoundly due to the colonization of the Seward Peninsula. It is important to note that the *Kigiqtaamiut* have never been "locked in a historical vacuum."[108] *Kigiqtaamiut* people made and continue to make selective changes in cultural traditions, social habits, and technological use as history unfolds and new situations arise. For example, snow machines and hunting rifles have been successfully incorporated into an already robust hunting tradition. It is also clear, however, that for Alaska Native peoples, the last one hundred years have been characterized by a history of outsiders imposing belief structures through powerful incentive programs, forced schooling, infrastructure development, economic giving and taking, and other mechanisms. These methods of intrusion have led some Alaska Native peoples and some scholars to identify the colonial period in rural Alaska as a period of genocide. While Alaska Native peoples were never passive observers of their own history, the last one hundred years have been marked by a violent imposition of ideological, political, and material consequences that we have come to understand as colonialism.

Federal development began in Shishmaref with the construction of a post office in 1901, a government school in 1906, and a Lutheran mission in 1930.[109] The convergence of education and missionization became an explicit goal of the US government following the end of the Indian wars and as the reservation systems became the standard-bearer of indigenous affairs. "The use of missionaries in dealing with American Indians involved the objectives of wholesale cultural change and assimilation into American society— principally through formal education commencing in 1871."[110] Alaska became a US civil and judicial district in 1884, making way for education policies to be carried out under the jurisdiction of the Secretary of the Interior[111] shortly after a formal push to handle the "American Indian problem" through policies of education and civilization instead of removal. This project of civilization was often carried out pragmatically with infrastructure development and legislative mandates.

Because of its absurdity, I love the quote that opens this chapter, taken from a missionary in Arctic Canada. "The old native home, from its very character, was a hot-bed of vice." This sentiment demonstrates for me the conflation, in the eyes of colonial powers, of infrastructure and cultural institutions. It demonstrates how development was actually a method of imparting Western institutions into rural Alaska and how the eradication of traditional infrastructure was a way to eradicate traditional institutions—including traditional and cultural adaptation strategies that had been developed over millennia for dealing with the local landscape. Obviously traditional sod houses were not "hot beds of vice"; what they may have been is sophisticated and resilient infrastructure, which was subsequently replaced by inefficient, more maladaptive infrastructure.

The most looming figure in the history of colonial institutions on the Seward Peninsula is Sheldon Jackson. Jackson promoted the trifecta of colonial planning: education, missionization, and industrialization. Jackson was appointed general agent for education in Alaska in 1885 and, despite having very little experience in rural Alaska, was fundamental in its infrastructure and colonial development. Federal funding at the end of the nineteenth century was insufficient to build and staff extensive school and church facilities throughout rural Alaska, so Jackson relied heavily on donations from Christian women's groups.[112] Whether because of the need to raise funds or due to character and personal conviction, Jackson was prone to hyperbole and often described rural Alaska in grossly exaggerated conditions of squalor, poverty, and oppression—an impression that has proven difficult to shake, even today.

For the women's church groups, he repeatedly spread the notion that Alaska Native women were considered exploitable property by their husbands. As an excuse to bring in domestic reindeer herds (a third wheel of civilization: industry), he declared widespread famine throughout the Seward Peninsula at the end of the twentieth century. The reality of this starvation period is controversial in the research.[113] Notably, Jackson made the argument that the population at *Kigitaq* had fallen from a height of two thousand people to a measly eighty people. This was most likely a misinterpretation of the explorer Fredrick Beechy's estimated population of the entire Seward Peninsula coast and a "fact" that played handily into the narrative Jackson was trying to perpetuate. Jackson's tale begins an interesting history of hyperbolized threat from outsiders for the purposes of aid and infrastructure development in Shishmaref and throughout rural Alaska.

Infrastructure development through schools and missions, and policies that required children to participate in school programs, resulted in the consolidation of smaller overwinter villages to the centralized location of Shishmaref. Settlements at Ikpik and Cape Espenberg gradually closed and the population of the old settlement of *Kigiqtaq* grew steadily over the next one hundred years.[114]

In my interviews, there are a variety of reactions to the development of the school and the church and their influence on Shishmaref today. Some people see forced policies requiring school attendance as the catalyst for the consolidation of family groups into the larger village. Others say that their elders knew the school was going to be important to future generations, so they chose to settle on the island. Most people express genuine pain when discussing how schools forbade speaking Iñupiat and in some cases imparted corporeal punishment to children who did so—resulting in the loss of the language for most people under the age of forty. There are mixed reactions from people I spoke with regarding missionaries. Christianity in Alaska Native communities is a complex spiritual system that incorporates new and old beliefs in different ways,[115] and churches are a central organizing feature of the community and social life for many people. The Lutheran Church in Shishmaref is a unique and meaningful place today and is a location where Iñupiat songs are sung and other cultural revitalization efforts are rooted. Going to church in Shishmaref can be a powerful tribute to the spiritual fortitude and grace of the community. What development has become in rural Alaska and the co-opting of colonial space for local use is the subject of another book.

For the purposes of this story, what is important is that the school came *before* sedentarization and the consolidation of family groups, and that these development decisions were made by outsiders who did not have extensive knowledge of the ecology of the region. Even when Shishmaref interviewees discuss sedentarization as a result of the wisdom of elders, it is always wisdom that came *after* the school was built. The original school infrastructure has been seen, in all of my interviews, as a product of outside decision-making.

MODERN INFRASTRUCTURE

While no longer the original buildings, the church and the school are still the largest pieces of infrastructure in Shishmaref today and are consequently where people gather for major events, from celebrations and funeral events to Christmas parties and athletic games. They are also evacuation centers for a major flooding event. While we know when the first framed-building school, post office, and church were built in Shishmaref, the transition from sod house to framed house for residents seems to have been gradual. In 1919, a schoolteacher responding to the influenza epidemic of 1918 explained that in the village there were daily inspections of people and "igloos."[116] Presumably, the igloos he is referring to are subterranean sod houses. He makes no mention of framed houses being checked by the nurse during the epidemic.

A 1920 transitional house made of driftwood—something between traditional sod houses in *Kigiqtaq* and framed houses already used at the way station in Deering—is described in Ellanna and Sherrod (2004), from oral histories taken with Gideon Kahlook Barr Sr. in 1991. These houses were located at *Ublasaun*, a village used by reindeer herders near Shishmaref following the importation of the reindeer herds to the area by Jackson:

> In 1920, from the exterior, the village resembled several small hills with protruding smokestacks. A small skylight made of the translucent stomachs of walrus or of glass was set in the apex of each of these sod-covered mounds—this skylight being large enough to emit light but small enough, hopefully, to deter the raiding paws of a polar bear.
>
> Gideon remembers that the family's 10-by-18-foot house at *Ublasaun* was constructed of driftwood. The small amount of scrap lumber available at Cape Espenberg was used to build the single bed for Thomas and Emily. Gideon and his siblings slept on the floor.[117]

Gideon remembers the conversion from a seal-oil lamp heating system to a cast iron stove as coinciding with the transition to this transitional type of housing structure. These intermediate housing structures stood more upright than traditional sod houses and were more dependent on outside materials such as glass, stove fixtures, and eventually lumber if sufficient driftwood was inaccessible. Transitional houses were not framed, were rounded at the top, and the outside construction was made with mostly locally available materials, including sod.

Today, most houses in Shishmaref are framed houses and are products of the federal government's Department of Housing and Urban Development (HUD). The HUD agency was formally authorized by the United States Housing Act of 1937. It wasn't until the 1960s that HUD began to prioritize American Indians as recipients of federal funds to promote home ownership. In 1969 the American Indian occupation of Alcatraz brought about increasing attention to the poor conditions of reservations across the United States, and in response, "President Nixon announced a new Indian housing initiative under which the federal government committed to the construction of thirty thousand new Indian housing units over five years. Alaska Senator Ted Stevens was influential in having HUD assign six thousand of the units to meet the housing needs of Alaska Natives."[118] Many older houses in Shishmaref date to this era, and at least two of my friends in Shishmaref live in homes that were previously their grandparents'.

In 1996, the HUD programs that were particularly aimed at providing low-income housing to Alaska Native and American Indian populations were consolidated and reorganized into the Native American Housing Assistance and Self-Determination Act (NAHASDA). This new legislation provides community block grants that are distributed through fourteen regional housing authorities, including the Bering Strait regional housing authority that serves Shishmaref.

OVERCROWDING AND DETERIORATING INFRASTRUCTURE

Today, housing in Shishmaref is insufficient for the population, and most houses do not have running water or sewer infrastructure. The lack of space and modern convenience is a contentious issue in Shishmaref. Many people I interviewed consider the housing shortage, overcrowded housing, and the lack of piped water and sewage as health issues. Epidemiological studies confirm that rates of gastrointestinal illness are substantially higher in communities without running water.[119] These health issues can be compounded by the outstandingly high cost of heating a framed house in the winter—which can run to over a thousand dollars a month. Shishmaref, unlike an increasing number of rural Alaska Native villages, does not have critical infrastructure development that includes a new health clinic, piped water, and significant new housing. The lack of modern infrastructure is linked to the community decision to relocate. In most cases, infrastructure-development projects in rural villages are the result of a competitive grant system filed through the Denali Commission or other state and federal agencies. Because Shishmaref has expressed the intent (through a vote) to relocate, it is an undesirable location for investment, and community infrastructure development has been minimal since 2002.

New houses are rare, and finding land on the island that is on sufficiently high ground for new houses is a challenge. Small lakes on the island that were used as water sources have been filled in to make space, yet the population continues to grow without adequate housing facilities. Multigenerational family homes, with up to twelve people living in a three-bedroom house are common in Shishmaref. The following interview points out that Shishmaref has received significantly fewer infrastructure developments compared to other villages in the region since the vote to relocate.

> If we're to remain here on the island, a lot of our grants that we apply for to expand our community public buildings like multi-purpose building or elders/youth facility, like a rec center, a bigger school, a bigger clinic—that's not possible because our island's too small and it's going to get smaller.[120]

When I asked what this resident would like to see in a new village, she responded:

> Running water, and just the fact that, you know, our community would finally be granted new buildings that we apply for so we don't have to live in these third world conditions. Be civilized like everybody else. To be provided services like any other community.[121]

I believe the lack of modern conveniences and housing is leading to the resettlement of talented, educated young *Kigiqtaamiut* to other cities. Outmigration is likely to increase if educated men and women who are poised to become local leaders are forced to live in overcrowded conditions. This is especially true for returning students with bachelor's degrees, jobs, and money to pay for apartments or houses—but without the infrastructure available on which to spend their money. I saw two exceptional young leaders move out of the village while I was there, and at least one expressed that this was directly tied to the lack of conveniences and overcrowding.

SEAWALLS AND SHORELINE STABILIZATION

Another critical piece of infrastructure and a critical expense for the community and the state has been shoreline stabilization. Shoreline stabilization was needed to protect framed infrastructure almost as soon as permanent settlement of the island became standard for *Tapqagmiut* people. Early development by missionaries began in 1901, and by 1940 seawalls and other revetment projects were already necessary to protect early development, but by that time the early missionaries who had established infrastructure on the island in the first place were gone. The *Kigiqtaamiut* were left to deal with high water that now, for the first time, felt like flooding.

According to the Army Corps of Engineers, the estimated cost of erosion control in Shishmaref to date is $9.5 million.[122] The lifetime costs of some revetment projects are estimated by Dr. Owen Mason to be up to $260 million. The first shoreline protection plan placed a series of fifty-five-gallon drums at the north side of the landing strip. The 1973 storms eradicated this effort. During the 1973 storm, more than fifty thousand sandbags were used to stabilize the bluffs located on the northwestern side of the island. These may have successfully prevented erosion during a large storm in 1974, but they were broken by ice in subsequent years. In 1982 a cement block revetment was constructed for one hundred meters along the bluffs, but it failed during the first big storm, within a year of its construction. In order to prevent continued erosion and dissipate wave action, residents pushed trucks, other vehicles, and old equipment over the shoreline. The following is a summary of seawall construction since 2004:

> In 2004, the BIA installed 200 feet of shoreline protection along the shoreline near the Native store. In 2005, the Corps installed 230 feet of protection connecting to the BIA project, extending to the east to protect the Shishmaref School. Also in 2005, the community of Shishmaref installed about 250 feet of protection extending to the east from the Corps project.[123]

All seawall and revetment efforts to date have been put in place to protect critical infrastructure. There are no efforts to protect the northwest side of the island, where most residents have racks and equipment that are used to butcher, dry, and put away subsistence harvests, especially black *ugruk* meat. This leads to the loss every year of traditional technology and equipment; for example, a storm in 2013 took out thirty to forty feet of land near the drying racks. The state does not feel these things are ideologically and economically necessary to protect. Below is a summary of the history of seawalls given by resident John Sinnok:

> EM: Do you want to relocate?
>
> JS: Yep. I can't see any other reason why we shouldn't. Like I just told you, the way that the village has been eroding, they've put rocks right in front of the village, but on the west side is where we have our racks to dry our meat, and my wife and I for the last three years, our racks are about maybe twenty feet. Every year for three years in a row we've had to replace them all. We've been three years in a row. We've lost at least sixty feet right there, our racks. If they don't save that part all of that is going to erode and there's just going to be this tiny village. And how much longer will those rocks stay? They've tried. In the 70s or sometime around there, they've tried to put a whole bunch of fifty-five-gallon drums welded together right along that beach, right along that land. Those stayed for a while, but they all sunk. Few years later they tried the sand gabions. I think those are below my mom's house. Used to be right by the Native store. Right there is used to be a long gentle slope and long flat land right there when we were kids. All that eroded. Then they put those gabions. I think that lasted for twenty-five years. But then they didn't do maintenance and it started eroding behind it. Couldn't keep the erosion away anymore. Gabions were bags of sand inside of wire. After those they tried cement blocks, going quite a ways under the sand. Those cement blocks—they started to topple right away. People have been putting their old trucks and stuff, and they sink right away. Anything that's not sand sinks.[124]

The Army Corps of Engineers also acknowledges the limited effects of the seawall projects: "All efforts to arrest the erosion have been unsuccessful for other than short periods of time."[125] The Shishmaref barrier island chain is a series of dunes built up in the sea. Today boulders and rocks are barged in to prevent this sand from washing away. Everyone I speak to understands this as a temporary solution, including residents of Shishmaref. "The seawall is just buying us a few more years. It won't last forever; we've got fine sand out there. Those rocks are going to sink eventually. Those are pretty big boulders."[126]

Shoreline stabilization offers very good short-term protection for critical infrastructure. What shoreline stabilization promotes is protection; what it discourages is flexibility. The history of infrastructure in Shishmaref reveals a gradual decrease in flexibility to weather and climate conditions over time, which corresponds to an increase of exposure and risk.

An illuminating passage by Owen Mason, published in a 2006 conference proceedings, states:

> About ten years ago, the State of Alaska sent me to Shishmaref to examine various alternative relocation sites, all on the mainland. In addition to this task, I considered the means available to remain on the barrier island chain. With some *flexible* engineering such as moveable structures and dune trapping devices (plants, fences, matting, etc.), I suggested that Shishmaref residents could remain in sync with the barrier or groom a nearby island for future settlement. The approach favored in the last 10 years has been the opposite: increasing hard stabilization, with the rocks larger and the lateral distance subject to seawalls longer. Further, the height of the wall is still far below the maximum storm surge limit, for reasons that I do not understand. [my emphasis][127]

The focus on building *flexibility* is striking in this passage. Moveable structures were essential parts of Iñupiat adaptive systems, and flexibility has been recommended by Mason, the state's leading archeologist in Shishmaref. In spite of these recommendations, the Corps' main objective appears to have been to keep the shore from moving and thereby to protect framed houses (purchased through federal grant and loan programs) as well as other critical infrastructure (as defined by the state).

CULTURAL VALUES AND INFRASTRUCTURE TRAPS: "WE LIVE HERE BECAUSE SHISHMAREF IS A GOOD PLACE TO HUNT AND BECAUSE THEY BUILT A BIA SCHOOL"

This brings us back to the larger question: Is the risk posed to Shishmaref the product of climate change or the product of a history of development that ignored local knowledge and removed local adaptation strategies? Climatic changes are not insignificant in the Shishmaref case study and these changes contribute substantially to the need to relocate. However, the simple equation that anthropogenic climate change = erosion = relocation is not an accurate analysis of this complex socioecological system.

As discussed above, like most hunter-gatherer societies, the *Tapqagmiut* were highly mobile before they were influenced by the presence of colonial institutions such as the mission and the school. This high mobility was linked to the ease with which infrastructure and other aspects of material life could be moved quickly. Before sedentarization, *Tapqagmiut* people were therefore able to make split-second decisions in response to changing weather conditions. High mobility and flexibility around weather was a successful adaptation strategy against flooding.

Building infrastructure was a key component to bringing education and Christianity into the Bering Strait region. The ideologies of education and worship are embedded in and expressed by infrastructure projects. Sheldon Jackson, the general agent for education in 1885, saw infrastructure and ideology as so interlinked that he raised money from outside federal streams to build schools and churches on the Seward Peninsula. This infrastructure project continued and expanded to include installing prefabricated houses, filling in island lakes to make room for new houses, and building a new, modern school in the late 1970s. Residents are aware of the irrefutable link between the first school that was built, moving to the island permanently, and the subsequent loss of flexibility to relocate easily. This is why Tony Weyiouanna said that "no one's talking about why we're here in the first place" in response to a question about contemporary relocation. Creating fixed, sedentary indigenous residents has been a strategy of the United States since the American Indian reservation project began and has been a goal of nation states all over the world.[128]

"Old Shishmaref," or the old village at *Kigitaq*, had been seasonally inhabited for five hundred years; high mobility and flexibility to storms provided an adaptation strategy for residents. Decisions regarding infrastructure development in the past remain somewhat cloudy; however, it is clear that Shishmaref residents and their ancestors made decisions that were grounded in an understanding of local ecologies. Sheldon Jackson, despite his great influence on early colonial development, had none of this localized ecological knowledge. In fact, some scholars claim that Jackson knew very little about rural Alaska at all, as the following passage suggests:

> Dr. Jackson had been credited with a profound knowledge of Alaska. This is a great exaggeration, for at best his knowledge was very superficial. In fact, it was his ignorance of the physical conditions in the Northland and of its people which led him to make many egregious blunders of administration. Another factor coupled with this was Jackson's fondness for sensational statements, no doubt in part developed as necessary to the propaganda to which he devoted most of his life.[129]

There is a strong indication that local knowledge was passed over in favor of outsider knowledge in early development decisions. The consequence of this is that permanent villages were built in vulnerable locations without adequate methods of coping with hazards.

Rural Iñupiat villages today have extraordinarily small populations—making the cost of modern protection from hazards seem disproportionately high per capita. Rural populations often do not have equitable access to typical Western strategies of adaptation that are highly reliant on technological intervention—like the levee system in New Orleans, for example. Nor do rural indigenous populations have access to traditional strategies of adaptation such as high mobility because of permanent infrastructure, which acts as the medium of modern and colonial institutions. Local adaptation strategies to an inevitably dynamic coast were lost without adequate new adaptation strategies to replace them.

This research predicts that the colonial development model will be more vulnerable to climatic changes and other ecological risks because development did not incorporate long-term, local ecological knowledge. This can be tested. If circumstantial, then indigenous communities should be no more exposed to flooding hazards linked to climate change than other communities. If, on the other hand, indigenous communities are more prone to live in locations that are exposed to repetitive hazards, then this historical understanding of vulnerability as a product of early colonial development may have important predictive power. In Alaska, 184 out of 213 (86 percent) Alaska Native villages experience problems with erosion and flooding.[130]

My research suggests why that might be the case.

Chapter Five

Finding a Way Forward: Trust, Distrust, and Alaska Native Relocation Planning in the Twenty-first Century

We wanted to make it, as much as we could, a local priority, from our local perspective with our cultural values—we want our village and our residents to be the actual people to be in charge of the relocation.

— Interview with Richard Kuzuguk, September 24, 2009.[131]

The Army Corps of Engineers predicted in 2006 that the village of Shishmaref had between fifteen to twenty years before continued erosion and habitual flooding would make the island uninhabitable. Today, relocating the Shishmaref community is still seen by most people as the only viable long-term solution for the village. Whereas in the past mobility was a part of the social fabric of life, today being on the move is more challenging, for reasons extensively discussed in chapter four. The millions of dollars of infrastructure that has been built in Shishmaref since 1901 now requires millions of dollars to be reconstructed or relocated. If a planned relocation doesn't materialize before a major disaster, Shishmaref residents risk death, loss of property, evacuation, diaspora, social disarticulation, increased landlessness, and increased poverty associated with the cascading outcomes of a flooding disaster followed by forced displacement.

So what is happening now? For the last forty years, Shishmaref residents have been engaged with government agencies to plan an organized relocation. This process has been an up-and-down cycle of interest and apathy, and it intersects with other aspects of village life in Alaska, such as rural development, changing demographics, and changing political structures and political will.

A TIMELINE OF RELOCATION PLANNING

Modern efforts to relocate the community began in the 1970s. Large storms in 1973 and 1974 made the island seem precarious and seriously threatened the safety of residents. In 1974, the Division of Community and Regional Affairs (DCRA) released a report on the Shishmaref relocation effort after a severe fall storm led to extensive damage on the island.[132] At that time there was prodigious planning by local residents to examine relocation options, and meetings between government agency representatives and local leaders established baseline costs and strategies for relocation. These plans did not come to fruition. The estimated cost of relocation in 1974 was placed at $1 million[133] for a complete relocation of the village. Today cost estimates from the Army Corps of Engineers range from $100 to $200 million.

Local residents say that in 1974 the majority of residents voted to remain on the island and not relocate. Percy Nayokpuk was in charge of these discussions in the early 1970s and still has copies of letters exchanged between the DCRA and himself that discuss possible relocation and timelines for relocation. Nayokpuk reports that the local decision to remain on the island was influenced by two factors. First, residents believed that shoreline stabilization would control erosion. Second, and even more influential to the vote, was that Shishmaref had become first in line to receive new state-funded construction in the form of a school. A vote to relocate would have removed the community from this list.

Official relocation planning then stalled for two decades, even though the storms did not. In 1988 a flooding event in Shishmaref was declared a state disaster, and a federally declared disaster occurred following a storm in 1997. These twenty years proved that seawalls and revetment projects would not provide adequate protection from flooding indefinitely, as discussed in detail in the previous chapter. A series of seawalls failed in Shishmaref over this time period, including a major cement-block revetment that was begun in 1982 and failed in the first major storm following its completion. Let me say that again: in 1982 a major cement-block revetment failed during the *first major storm* following its completion. In 2001 another significant storm flooded the village. It was after this storm that the community formally established the Shishmaref Erosion and Relocation Coalition. The coalition is a formal power-sharing agreement between the Native Village of Shishmaref, the City of Shishmaref, and the Shishmaref Native Corporation.

The 2001 storm was significant enough to garner statewide public attention, media attention, and the attention of regional leaders. Then-governor of Alaska Tony Knowles issued an administrative order declaring that "not doing anything [in Shishmaref] would pose an imminent and continuing threat that justified the State taking action to provide some kind of protective measure along the shoreline of Shishmaref."[134]

The attention and energy surrounding relocation was significant enough by 2001 and 2002 that residents considered relocation to be forthcoming. An entry on the Shishmaref Relocation Coalition's timeline of important events reads: **April 30, 2009: Move to new site is complete.** Interviews with residents reveal the high expectations and subsequent letdown of this period of planning. One woman, asked about her feelings following the vote in 2002, said, "We were like, cool, everybody want to go, people are going to get funding. We had all these high hopes, you know? We thought it was just going to happen, but in reality that does not happen at all. You know, we're still here."[135] In another interview, a resident commented, "At that time [2002] we were led to believe, we had a chance at that time [to relocate], but not understanding what the total process was at the legislative end was hard to picture.[136]

After the 2001 storm some Shishmaref community members embraced "getting the story out" as a strategy to garner political attention. Before 2001, when Tony Weyiouanna lobbied state and federal lawmakers for erosion protection and relocation funding, he had a difficult time getting anyone to pay attention. With over two hundred small Native villages in rural Alaska, and with the perpetual challenges of small-scale development and other rural needs, environmental threats were considered an outlier and a low-priority concern of the legislature.

In the early 2000s the relocation of rural Alaskan villages driven by climate change was an invisible issue to everyone except the flooding victims themselves. In the notes from the Immediate Action Working Group, board members acknowledge, "These problems [flooding and erosion leading to relocation], which primarily affect small, isolated communities, *are difficult to address and due to this are easily ignored* [my emphasis]."[137]

Fighting apathy and frustrated with being ignored, Weyiouanna decided that media attention may be a method of garnering public support for the flooding problems in Shishmaref. In 2002, a *People* magazine reporter phoned Weyiouanna in his office. The reporter had an ultimatum: He told Weyouanna to convince him to come to Shishmaref, or he was going to Tuvalu for a climate change story.

"I looked up Tuvalu on the internet," says Weyiouanna, "and saw that he could be sitting in shorts drinking a margarita." Or, he could come to Shishmaref, the relatively desolate island in the middle of the Chukchi Sea. But, Weyiouanna says, "What I knew we had was culture."

Weyiouanna and others were successful in cultivating media attention, but it has been a double-edged sword. Once the floodgates opened, it was hard to keep reporters out. A Shishmaref Relocation Coalition Newsletter from 2006 reports that sixty-four news organizations had visited the community since 2002. The partial list of news and documentary organizations that have visited Shishmaref for climate change pieces includes, but is not limited to, the *New York Times*, the National Film Board of Canada, the Associated Press, Reuters, *People* magazine, Earthwatch Radio, Global Create (Japan), *National Geographic* magazine, Maison Radio (Canada), Viverra Films (Holland), the *New Yorker*, The Weather Channel, BBC, *Time* magazine, TV Asahi (Japan), ABC News, French Daily Liberation, HBO, the Norwegian Broadcasting Corporation, Thalassa (French television), HD Net TV, National Public Radio, the German TV network ZDF, Svenska Dagbladet (Sweden), and CBS News.

My impression is that many Shishmaref residents feel uncomfortable with the attention. Locally, there is a high priority on the accuracy of information, which includes resisting exaggeration. Shishmaref residents exert these social norms in interviews with news organizations, but when the interviews are embedded in media contexts, they are often bracketed by catastrophic language. In the following news pieces about Shishmaref, we can see the divergence between local language and the journalists' language. From the Associated Press:

> Thousands of years ago, hungry nomads chased caribou here across a now-lost land bridge from Siberia, just 100 miles away. Many scientists believe those nomads became the first Americans. Now their descendants are about to become global warming **refugees** [original emphasis]. Their village is about to be swallowed up by the sea. . . . "We have no room left here," said 43-year-old Tony Weyiouanna. "I have to think about my grandchildren. We need to move."[138]

As another example:

> When the arctic winds howl and angry waves pummel the shore of this Iñupiat Eskimo village, Shelton and Clara Kokeok fear that their house, already at the edge of the Earth, finally may plunge into the gray sea below. . . . "The land is going away," said Shelton Kokeok, 65, whose home is on the tip of a bluff that's been melting in part because of climate change. "I think it's going to vanish one of these days."[139]

In the next excerpt, the author comments explicitly that Weyiouanna is unemotional about his statements and that Weyiouanna speaks with "the indifference of an engineer":

> "I don't think we have much choice now," he tells me on the eve of the new ballot. "Some might vote no—people so tied to the island they don't want to leave. We'll just have to make adjustments." Like a wholesale migration to the mainland, an adjustment he discusses with the indifference of an engineer, not someone who's lived here all his life.[140]

In these examples, the quotations from residents are: "I don't think we have much of a choice now," "The land is going away," "I think it's going to vanish one of these days," "There's no room left," and "we need to move." These are all experiential statements of erosion that are grounded in physical realities. The journalists insert the catastrophic language of "wholesale migration," "angry waves," "edge of the earth," "and swallowed by the sea."

Carol Farbotko and Heather Lazrus have pointed out that climate change refugees have been used as "ventriloquists for climate change narratives" by outsiders.[141] Similarly to discourses by Shishmaref residents, Tuvaluan narratives about relocation and flooding are often less catastrophic and are framed by ideas of global citizenry and human rights as opposed to being framed by helplessness and victimization.

Publicity regarding the challenges faced by Shishmaref residents is not the same thing as Shishmaref residents having an opportunity to frame and express their own experiences to a wider public. Media attention, however, did precede and may have helped secure government attention for climate change concerns, including flooding and erosion. Shortly after the media picked up "the Shishmaref story," a number of government agencies began looking into climate change issues along the Alaskan coast.

In 2004 a report from the US Government Accounting Office (GAO) identified the extent of flooding problems in rural Alaska. The report was mostly concerned with articulating policy obstacles to access federal funding by Alaska rural villages. It was the scope of the problem identified by the report, however, that caught people's attention—86 percent of Alaska Native villages facing flooding problems wasn't a disaster, it was a permanent condition and an example of a permanently changing environment. Alaska rural villages, especially Shishmaref, were a harbinger of climate-induced displacement.

Two years later, the Army Corps of Engineers looked into the cost of solving the problem. In 2006, the Corps published a research inquiry into possible solutions for Shishmaref relocation. Three possible scenarios and their outcomes and costs were described: (a) relocate to the mainland and reconstruct village infrastructure from scratch, (b) relocate residents to the regional centers of either Nome or Kotzebue, or (c) take no action. The conclusion of this report reiterated that relocation to a discrete village on the mainland (outside of Nome or Kotzebue) was the only feasible long-term solution for Shishmaref residents. This conclusion was not binding, however. A 2009 report from the Immediate Action Working Group (described below) includes the "do nothing" and colocation possibilities in their report for Shishmaref.

One problem, identified early by state and federal agencies, was that no single government agency had the authority, mandate, or funding mechanism to oversee the complex coordination of community relocation. Relocation requires the development of roads, airports, schools, barge landings, houses and the institutional organization for these things to occur in tandem. Currently these myriad projects are all handled by different agencies—each of which has their own budgetary cycle and funding streams. In 2007, in an attempt to promote coordination, then-Governor Sarah Palin established the Alaska Climate Change Sub-Cabinet.

The Sub-Cabinet was then subdivided into working groups. The Immediate Action Working Group (IAWG) was charged with recommending strategies to avoid disasters in places and areas that were in imminent risk of disaster[142] and was made up of high-level representatives from the US Army Corps of Engineers; the Department of Commerce, Community, and Economic Development; the Department of Natural Resources; the Department of Transportation and Public Facilities; the Denali Commission; the Alaska Municipal League; the Alaska State Legislative and Budget Committee; the Alaska Division of Homeland Security; National Oceanic and Atmospheric Administration; Alaska Tribal Health Consortium; the Environmental Protection Agency; and the US Economic Development Administration.

It was this body, led unofficially by Sally Cox, that tackled the relocation issue. The IAWG identified six communities that were the most significantly affected by climate change and that needed immediate attention, including the possibility of relocation. These communities were Kivalina, Newtok, Shaktoolik, Shishmaref, Unalakleet, and Koyukuk.

THE REVOLVING-DOOR THEORY

When I first began my research in Shishmaref, a preliminary finding was that high turnover rates among state and federal bureaucrats working on relocation, as well as short-term budgets for relocation and risk-mitigation intervention, created a situation in which Shishmaref residents were constantly dealing with agency workers who had no background knowledge and no historical awareness of local protocol, previous relocation studies, or previous government efforts—as I discussed above. The revolving door theory meant that Shishmaref residents were always working with people who were blind to relocation history and were reinventing the wheel. In Shishmaref, local, low-level bureaucrats are more stable, and institutional memory is longer because turnover is often among relatives or friends. From 2007 to 2009, however, I began to reconsider the revolving-door theory of bureaucratic workers in Shishmaref who worked on risk and relocation. The IAWG seemed committed and stable. The group had significant funding and was actively seeking to add new villages to the list of communities they worked with and served. It seemed reasonable to believe that the IAWG would become a clearinghouse committee to handle relocation issues. The IAWG reports were trying explicitly to develop protocol and step-by-step planning for at-risk communities who need to relocate.

However, 2009 was the last time the IAWG produced a report, and the committee has since disbanded. The disbandment of the IAWG demonstrates what Shishmaref residents have often pointed out: There is no cohesive planning, and, as new iterations of help and strategizing committees arrive, the community and planning phases have to start all over again. This is often discussed in Shishmaref as "another study being done." Tommy Obruk commented, "you know that kind of slows them down, the studies. Government always works real slow to do the studies."[143]

A new organization directed by the Department of Commerce, Community and Economic Affairs and the DCRA, called the Alaska Climate Change Impact Mitigation Program (ACCIMP), has followed in the footsteps of the IAWG. The ACCIMP is attempting to set protocol for villages and village leaders to follow. This is an important step in the relocation process. Without clear steps in the relocation process, local and state leaders are inefficient at streamlining funding. I feel a surge of hope with this new development—similar to when the IAWG was at its most active. The two contact personnel listed on the ACCIMP website, Sally Cox and Erik O'Brien, are both individuals who have worked on relocation issues since the beginning of this research, though mostly in Newtok. Even so, this is a change from the IAWG, which mostly had high-level bureaucrats as board members.

In any case, the bureaucratic organization has undergone yet another iteration, and people in Shishmaref are still waiting.

THE CURRENT STATE OF AFFAIRS

In Shishmaref today, leaders are more hesitant to discuss relocation than they have been in the past, and some are making renewed claims that the community is going to stay on the island for the foreseeable future. While the majority of residents have voted to relocate, a set of new, vocal leaders have noted that discussing relocation has inhibited development without actually securing a plan to relocate. Shishmaref has recently had some upgrades to their medical clinic (though they have not received a new clinic) and there has been some new housing. This is the most substantial development in a decade for the underserved community. While relocation remains the only long-term solution, the continued failures for relocation planning on the part of state and federal agencies have left communities bereft of options and increasingly underserved. Talking about the need for relocation in and of itself has become a liability for villages at risk. The situation is increasingly compromised.

THE PRAGMATICS OF PLANNING

Tracing the timeline of relocation planning in Shishmaref does not, in any way, capture what it feels like to try to plan relocation. The IAWG was absolutely a well-intentioned group charged with tackling a bureaucratic organizational nightmare, but the Shishmaref residents' dealings with the IAWG could be problematic and confounding. A lack of personal relationships between Shishmaref residents and IAWG members and a lack of historical awareness on the part of the IAWG members could lead to miscommunication.

The following is an excerpt from an IAWG meeting that highlights the actual pragmatics of relocation planning for the Shishmaref community:

> *May 17, 2010*
> *(reconstructed from field notes, the Immediate Action Working Group meeting agenda, interviews, and memory)*
>
> Board members of the Immediate Action Working Group (IAWG), a subdivision of the Alaska Governor's sub committee on climate change, meet to further discuss what to do about the increasing number of Alaska Native villages that are experiencing problems with erosion and flooding linked to anthropogenic climate change.

Today the board meets in Anchorage, Alaska, in an office building, around a conference table that is full of briefcases and computers. Someone is paid to take notes. These men and women representatives from state and federal agencies know one another, and before the meeting begins, they exchange pleasantries. At the center of the table is a conference telephone. No Alaska Native participants from affected rural communities are physically present at the meeting, but at least eight participate by phone.

Six hundred and four miles away, in Shishmaref, sit five community members in the basement of the local church. The IAWG has put up information in real time on the web concerning the agenda, but the Internet connection in Shishmaref is too slow to pull the agenda slides up as the Anchorage participants move through them. It is difficult to hear what the board members are talking about from a small speakerphone in Shishmaref.

To remedy this, Fred Eningowuk finds a karaoke machine from the basement and puts the microphone to the speaker so that the sound is now being broadcast over a makeshift sound system. This is marginally successful, but the sound wavers between static-infused mumbling and booming loudness, depending on who is talking in Anchorage and where they are seated with respect to the telephone.

I sit with Shishmaref participants in the church basement, incredulous at the fact that this simultaneously tedious, impossibly frustrating, and somehow hilarious episode of bureaucratic karaoke is the functional mechanism for avoiding catastrophic flooding and diaspora for the Iñupiat residents of this threatened community.

This meeting lasts for four hours, and the sound quality continues to make it difficult for everyone to follow—all the more so for the elders in the room, for whom English is a second language. By the end I am pained and exhausted and so uncomfortable in my chair that, even though I am trying to be still, I shift constantly in my seat. Shishmaref residents participating in the meeting are much more still in their chairs.

Finally, after waiting for hours while the board discusses criteria for adding new communities to the "imminent risk" list, it is time for Shishmaref residents to give their update on local concerns and progress to the board in Anchorage.

Eningowuk tells the group in Anchorage that Shishmaref needs help getting old, abandoned bulk-fuel containers from a nearby village into Shishmaref. The fuel containers in Shishmaref are eroding, but Shishmaref hasn't been eligible for new fuel containers, or most other state-sponsored infrastructure upgrades, since the village voted to relocate in 2001.

Instead, residents have used social networks to locate some abandoned tanks from the nearby village, but they do not have the transportation capacity to move them from one village to another, so they are asking for help from these people in

Anchorage who are the working government body charged with helping villages who need to relocate.

There is silence from the board in Anchorage.

Fuel containers and fuel container transportation do not fall under the mandate of the working group, so the group in Anchorage moves on without comment. They literally ignore Eningowuk's comment. Say "thank-you" and move on.

I feel extraordinarily embarrassed.

So there you have it: the participatory, bureaucratic mechanism through which a long-standing, indigenous community is supposed to plan their relocation.

Sitting in Shishmaref, the connections among climate change, environmental migration, and bulk-fuel containers are clear. Also clear are the links between funding streams, immediate risk, long-term risk, local poverty, poor internet connections, and the increasing reliance on outside decision-makers for aid and risk mitigation. I estimate that (at least) 90 percent of the four-hour meeting is taken up by government agency workers discussing amongst themselves disaster mitigation planning, interspersed with seemingly disconnected comments by Native leaders from rural Alaska via phone.

"Waste of time," says someone after the phone call is finished. "Waste of time."

This scene depicts the actual setting in which climate change, disasters, and relocation are managed. For me, experiences like this one highlight the way vulnerability to climate change can be compounded and routinized. Language barriers, technological differences, and the gap between local needs and state functioning are the real, everyday mechanisms that haunt disaster mitigation. In an indigenous context these interact with historical prejudice and colonialism in complex ways. In Shishmaref, distrust and lack of understanding between agencies and communities are real factors that challenge successful progress.

OBSTACLES TO RELOCATION: CLIMATE CHANGE, DISTRUST, AND PARTICIPATION

As might not be surprising, I found that Shishmaref residents distrusted the efficacy of government planning agencies and, in some cases, considered agency workers likely to misunderstand the experiences of Shishmaref residents.

As a supplementary methodology to more classical ethnography, I conducted a small survey (thirty adult respondents) in Shishmaref regarding attitudes about relocation and government planning. The most interesting results from the survey demonstrated that Shishmaref residents were highly concerned about climate change and were distrustful of government relocation strategies.

In the survey, respondents were asked to indicate the extent to which they agreed with the following three attitude statements on a scale ranging from 1 (strongly disagree) to 5 (strongly agree): (1) "I feel confident that Shishmaref will be relocated in a timely manner before a major disaster occurs"; (2) "It is clear to me which government agencies would fund relocation"; and (3) "Global warming, or climate change, is the greatest threat to Shishmaref's future."

> *Results:* On average, survey respondents strongly agreed that "climate change is the greatest threat to Shishmaref's future" ($M = 4.60$, $SD = .88$).

Despite high unemployment, inadequate housing, and other economic difficulties, climate change was at the forefront of people's minds as a pervasive and ever-present danger that will affect the future. While climate change is not the only driver of vulnerability, it is locally perceived as a major threat.

> *Results:* On average, survey respondents tended to disagree that "Shishmaref will be relocated in a timely manner before a major disaster occurs" ($M = 2.05$, $SD = .94$).

Although there was variation among residents, overall, those surveyed *did not believe* that Shishmaref would be relocated in an organized way before a major disaster occurred. In interviews and during conversations with friends, Shishmaref residents expressed fears about diaspora and expressed deep wishes to keep themselves and their families in *Kigiqtaamiut* traditional territory. I was surprised to find that so many people simultaneously believed that this was not likely to happen in an organized fashion. The survey captured this measurable sentiment: *Shishmaref residents do not have high confidence in government relocation-planning efforts.*

Perhaps most interestingly, results from the survey revealed a correlation between the lack of confidence that an organized relocation would occur before a major disaster and clear knowledge about which government agencies would fund relocation.

> *Results:* Survey respondents who *agreed* with the statement "I am clear about which government agencies would fund relocation" were likely to *disagree* with the statement "I believe that Shishmaref residents will be relocated in a timely manner before a major disaster" ($r(28) = -.47$, $p = .02$).

In other words, *awareness of bureaucratic processes predicted low confidence in bureaucratic processes.*

That is to say, the survey found that when respondents self-reported an awareness of the mechanisms of state and federal agency intervention, they were less likely to have confidence in government intervention to organize a planned relocation. One problem in the ongoing efforts to organize relocation is that residents do not believe that agency workers have an accurate appreciation for either the lifestyle of Alaska Native peoples or the extent of risk that Shishmaref residents experience.

THE PERSONAL AND PROFESSIONAL GAP IN RISK MANAGEMENT: WHAT GIVES KIM NINGEALOOK GRAY HAIRS

During an interview with Kim Ningealook,[144] I asked if relocation caused him any stress or frustration. "It's what's giving me gray hairs!" he said. Stella, Kim's wife, and I laughed and laughed—but there is a tension in Shishmaref between the stress of storms and relocation, on the one hand, and the joys of daily life in a close-knit community, on the other. In an interview with Steve Samuels, the principal of the Shishmaref school in 2010, he said, "[Relocation] doesn't ever seem to have a positive spin. It's almost always a negative thing. Not that people obsess about it or anything, but it does come up from time to time. . . . It seems to be a sad thing, in my experience."[145]

There are full lives being led in Shishmaref that have nothing to do with relocation, but persistent risks and negative feelings about relocation are pervasive. Researching disaster was a sad business and lent itself to difficult discussions—my experience in Shishmaref included asking people to discuss some things that they would rather not. I found that as an interview went to twenty-five or thirty minutes, sometimes to an hour or more, greater fears and concerns emerged, and people expressed more latent distresses. This interviewing was not easy for community members and should make researchers, agency workers, and journalists wary of overburdening or making unreasonable demands for information from at-risk communities.

There are three very common concerns in Shishmaref regarding risk and relocation. The first is that risks to the island are increasing—both from flooding and erosion and because the ice conditions are changing so rapidly. This especially presents challenges to hunting patterns and can create risky conditions for hunters. Table 5.1 lists representative responses from interview data and personal conversations regarding how residents experience changes on the landscape. In light of change and in light of the forty years of relocation planning without an outcome, residents can feel a sense of helplessness. This should not be overstated—in the following chapter I will discuss the resiliency and tenacity of Shishmaref residents in the face of dramatic changes—but present in interview data is also the feeling of despair, which is demonstrated in the first two responses in table 5.1.

TABLE 5.1 Interviews on Risk

Risk is increasing in Shishmaref linked to climate change	"It's [the island is] going to go away until there's nothing. It is global warming and it is mother nature that we can't help" (B.E.).
	"It feels like we're sitting on a big tub, like it's going to fill up with water. That's how it feels being on this island" (K.N.).
	"Biggest change is that climate change is playing such a big effect in our community, not only that the ice is thinner. The water's too close for hunting with snow machines" (R.K.).

TABLE 5.2   Interviews on Disaster and Diaspora

Disaster will occur (and lead to diaspora) before relocation can be organized	"I don't believe the political structure/process can do it [fund relocation]. It is too slow. It's always in the planning stages, but there's no funding for it. One day we are going to be evacuated" (R.K.).
	"To not act represents the annihilation of our community through dissemination" (Shishmaref Erosion and Relocation Committee).
	"Nothing's being done. Look, we're still here" (J.D.).
	"Just scared if we relocate we're going to have to move to different towns" (Y.M.).
	"We'll be scattered like refugees" (R.K.).
	"Most of the conversation that I hear around relocation, the people don't have a real positive feeling about it, not that they don't want to relocate, but that they don't think that there's a site that's viable" (S.S).

The second major area of concern is the fear of diaspora following a major flooding event, before an organized relocation occurs (see table 5.2). On the ground, a palpable fear is that disaster and evacuation will eventually cause the dissemination of the Shishmaref community into different villages, towns, and cities. Diaspora is a significant and extremely emotional topic. It is a fear that spans multiple generations. During an interview with three young hunters in their late teens and early twenties, they spoke of the scattering of the Shishmaref population as their biggest fear—similar to that of the elders and elder hunters I spoke with on the island. There is not a plan in place for what will happen after a major storm if the island becomes uninhabitable and residents are evacuated, but this scenario is salient among residents. The Shishmaref Erosion and Relocation Committee formally considers dissemination or diaspora of the population to be an annihilation of the community and an annihilation of the cultural integrity of residents.

The third concern in Shishmaref is that agency workers and other decision-makers do not really understand what Shishmaref residents go through—that these outsiders are uneducated about the history of Shishmaref relocation planning and that there is miscommunication and misunderstanding between agency workers and local residents. There is a sense in Shishmaref that if government workers could just experience a storm for themselves, then this experience would translate into forthcoming state and federal aid. Table 5.3 lists interview excerpts that describe this frustration.

Trying to negotiate your life, livelihood, and the survival of a subsistence tradition through bureaucratic mechanisms that are slow, repetitive, and unresponsive and are

TABLE 5.3   Interviews on Communication

Alienation and communication difficulties with bureaucratic agencies	"People say, we don't need to go to those meetings, they just go around and around. We won't move; we won't ever move" (A.K.).
	"They've got to see it to believe it" (Anonymous).
	"It's been, same every time. It's like, how come you guys are here again, saying the same stuff? We already heard this last time, you know?" (J.D.).
	"Let the federal agencies come here and experience a whole storm, not come for the day and leave. Let them be here two weeks, so they could see it for themselves, cause it always seems like they don't believe us" (J.S.).

staffed by people who live extremely different lives than you do seems like a special kind of hell.

People in Shishmaref are tired.

Many residents report that, starting in 2002, there have been multiple meetings a year to discuss relocation with the community, often sponsored and hosted by state and federal agency workers. For residents, these meetings are redundant. High turnover rates among agency workers lead to black holes of information regarding relocation. Each successive generation of outsiders is tasked with analyzing some aspect of relocation that differs only slightly from that explored by the agency worker who came before them. When a new agency worker or engineer or researcher comes in, residents must tell and retell the story of relocation planning from the beginning. As this process of educating outsiders becomes frustrating, community members participate less and less. Annie Kokeok, the Kawerak transportation coordinator in 2010, talked about how motivating residents to participate in the process again was a challenge; she observed that morale about the likelihood of moving was low.

New state and federal agency workers can misunderstand a lack of community participation as apathy. Agency workers today are devoted individuals who are dedicated to culturally appropriate solutions and have a stake in Shishmaref relocation, but personal concern is obscured by short-term participation and a lack of historical awareness. For example, in an early 2008 meeting with the IAWG, one observer (not a member of the IAWG) suggested just moving the school and thereby forcing residents to relocate in order to remain with their children. This is a strategy very similar to Sheldon Jackson's early twentieth-century planning—and a very sensitive issue to most people I know in rural Alaska. This top-down social engineering is what ultimately creates vulnerability in the first place and is a racist ideology that drove so much hurt and sorrow in indigenous communities. Hearing it reiterated in the twenty-first century was truly a heartbreaking moment for me in this project.

Local participation is cyclical and can go through waves of energy and exhaustion. A new generation of leaders is taking over in Shishmaref; in some cases, sons are literally taking over positions held by their mothers. As younger members of the community enter into office, there is renewed energy. However, any outsider working in Shishmaref (researcher, engineer, journalist, bureaucrat) should be vigilantly aware of the repetitiveness with which Shishmaref residents have told and retold, educated and reeducated outsiders about the history of relocation—and how often they have heard of new strategies and surveys being done before construction can begin. As Jennifer Demir said, "It's like: how come you guys are here again, saying the same stuff. We already heard this last time, you know?"[146]

Residents were most satisfied with the attention and aid given by the late Republican Senator Ted Stevens. This is an interesting political relationship, considering that most Alaska Native villages vote for Democratic candidates in state and national elections. Their satisfaction, I think, was linked to Senator Stevens's tenure as an Alaskan government servant and his personal visits to Shishmaref. One person told me, "The biggest help that ever came in and we had stuff done after the visit was Senator Ted Stevens. He *saw the erosion for himself* [my emphasis]."[147]

In 2002, I witnessed Senator Stevens engaging with a local community on the Bering Strait during a trip to Little Diomede. My overwhelming impression of this event was how well Senator Stevens related to and took seriously the concerns of Little Diomede residents. As the interview above suggests, the fact that Senator Stevens *saw the erosion for himself* was critically important.

As we've seen in the history of Shishmaref infrastructural and institutional development, outside developers, politicians, and decision-makers who ignored local knowledge helped create vulnerabilities to ecological shift in the first place. This may be part of the reason why Shishmaref residents today are quick to focus on personal and local experience as a key component to expertise in relocation issues. When outside planners, researchers, and stakeholders lack local experience and long tenure in the region, this can lead to misunderstandings between themselves and local residents.

The most obvious solution to these problems is clear protocol that places decision-making power in the hands of residents themselves. This desire was reiterated over and over again in interviews. One elder commented, "They come here and said pretty nice things about what they would do to help our village, but I read it, and it makes it where it's only their decision, and I don't think that's right."[148] In another interview, a resident responded, "We wanted to make it, as much as we could, a local priority, from our local perspective with our cultural values—we want our village and our residents to be the actual people to be *in charge of the relocation*."[149]

When political organization across scales is necessary, as it is and will be, then emphasis must be placed on local articulation of the problem and possible solutions. From a social scientific perspective, local decision-making power is not only an outcome of reducing political vulnerability but also may create an added layer of ecological resiliency because local planning capitalizes on local knowledge of ecological conditions.

POLICY CHALLENGES TO RELOCATION

The data presented in the previous section outlines some of the complex social challenges of planning an organized relocation. These include high turnover rates among agency workers and misunderstandings between agency workers and local residents. This section will discuss institutional and policy obstacles to relocation.

Current US policy on disasters makes an organized relocation problematic at best and impossible at worst. Rural Alaska Native villages serve as case studies for climate-change-driven environmental relocation in the United States precisely because they demonstrate the ways in which disaster governance is ill equipped to handle community-wide migrations, particularly as climate change creates new ecological norms.

FEMA is the primary political mechanism through which disasters are both responded to and mitigated in the United States. After a disaster, FEMA acts as an umbrella organization and has the power to simultaneously coordinate disparate agencies and infrastructure projects. FEMA is governed by the Stafford Act of 1988 (amended 2013), which, among other procedural amendments, outlines the goals of disaster recovery as promoting "recovery through rebuilding." As the Stafford Act states, "because disasters often disrupt the normal functioning of governments and communities, and adversely affect individuals and families with great severity; special measures, designed to assist the efforts of the affected States in expediting the rendering of aid, assistance, and emergency services, and the *reconstruction and rehabilitation* of devastated areas, are necessary" (Sec. 101[a(2)], my emphasis).

The Stafford Act sets rebuilding in place as an explicit goal of disaster response,[150] which is illogical in places like Shishmaref that are becoming increasingly uninhabitable due to increased exposure to flooding.

There is no corresponding agency for *preemptive* disaster planning or disaster mitigation in cases where erosion increases exposure to flooding hazards. Relocation planners, researchers, and *Kigiqtaamiut* advocates all recognize the organizational nightmare of attempting to coordinate multiple governmental agencies and their annual budgets to plan an organized, timely, community-wide relocation.[151] The effect is that every step (barge landing, airstrip survey, housing developments) must be funded and undertaken individually.

Relocation of individual households is possible in the United States under the Voluntary Buyout Program. The Voluntary Buyout Program can be carried out through four funding mechanisms: the Flood Mitigation Assistance (FMA) program, the Hazard Mitigation Grant Program (HMGP), the Pre-Disaster Mitigation (PDM) program, and the Severe Repetitive Loss (SRL) grant program. The Voluntary Buyout Program provides homeowners and business owners with the opportunity to sell properties using federal funding to supplement some (but typically not all) of the value of the properties. State and local agencies typically split costs of voluntary buyouts with the federal government at a rate of 25–75 percent share. Participation in the Voluntary Buyout Program is, as the name suggests, voluntary at the individual household level.

This program is not (as of yet) applicable to the Alaska case studies for two reasons. First, the matching requirements from local sources are impossible to meet given the cost of relocation and Shishmaref's population. Second, this policy is not equipped to handle a community-wide buyout, which would require new community-scale infrastructure, such as a new school, new medical facilities, and new roads; it has only been used for individual households relocating into preexisting towns and cities.[152]

The Voluntary Buyout Program has the additional challenge of being a secondary disaster mitigation strategy in the United States. Between the years 1993 and 2011, FEMA spent approximately $2 billion to buy back 37,707 high-risk properties.[153] Missouri, following the Great Flood of 1993, was the state most actively participating in the buyout program. In spite of the overwhelming disasters of Hurricane Katrina and Hurricane Sandy, Missouri remains the most active state in the program. The $2 billion in funding over eighteen years, while significant, pales in comparison to the Obama administration's request for $60.4 billion for recovery and rebuilding along the eastern seaboard after Hurricane Sandy or the $81 billion in property damage following Hurricane Katrina.[154]

Because there is no real policy protocol in place to handle relocations from Alaska Native villages, the outcome is that individual communities in Alaska are embracing different strategies and are having different challenges and different successes.

THE (RELATIVE) SUCCESS OF THE NEWTOK PLANNING GROUP

One example of the lack of organization and established protocol in relocations in Alaska is the fact that out of all the villages that need to relocate, one village, Newtok, seems to be proceeding rapidly where the others are not. To date, a barge landing and access road have been built and an evacuation shelter is being constructed at a new relocation site. Despite this initial progress, relocation planning in Newtok has also stalled.

The progress that has happened in Newtok is often attributed to the coordination among the Newtok Traditional Council, led by Newtok resident Stanley Tom, and state and federal agencies—together this group constitutes the Newtok Planning Group. The Newtok Planning Group was considered by the IAWG a "model for local, community, state and federal partnerships to address complex issues—the community planning efforts have enabled the community to advance its already innovative successes."[155] In at least one phone meeting for which I was present, an IAWG board member suggested that progress in Shishmaref was stalled because of the lack of local organization. In the 2009 report, the IAWG stated that Shishmaref had "community planning needs to coordinate with the various organizations to effectively plan for the needs of an entire community."[156]

The Shishmaref Erosion and Relocation Coalition was convened in 2001 (before the Newtok Planning Group), met regularly, and is representative of all the local political affiliations. I'm not sure why agency workers perceived a lack of local organization in Shishmaref, but it may have to do with the perception of local power. In Newtok, success is often attributed to Stanley Tom.[157] In Shishmaref there is not one individual, but a collective, that must be engaged. This requires state and federal agency workers to deal with multiple attitudes and opinions. The multiplicity of voices may challenge the image and stereotypes agency workers have of Alaska Native village politics and cultural norms.

Because multiple villages need to be relocated and because these relocations are very expensive, the progress of one village over other villages can be highly sensitive and could become politically and financially competitive. I have witnessed and recorded Shishmaref residents defusing this potentially sensitive and competitive situation among "at-risk" communities by consciously and explicitly identifying erosion and flooding as a collective struggle and by characterizing the success of another "at-risk" village as also a success for Shishmaref.

When asked about the impact of the IAWG on successfully helping with relocation, one young community leader responded, "Yeah, not necessarily for Shishmaref but for other communities. They're helping Shishmaref in a way too. If they're helping one community we get a little better edge on what we need to ask for."[158]

In a separate interview, an elder leader echoed these sentiments, saying "on that teleconference yesterday I noticed that some lady spoke that they were trying to get a relocation planner for Kivalina and Shishmaref. That was great, to help both villages, you know."[159]

SITE SELECTION

In Newtok, selecting a site has been a key decision that has sped development.[160] The Newtok Traditional Council prioritized good soil, lack of erosion, subsistence accessibility, barge accessibility, space for an airport, and not infringing on other villages' subsistence practices as key characteristics of a culturally appropriate site. These criteria are similar for Shishmaref residents—but problems may arise if one or more criteria need to be compromised.

In Shishmaref, possible relocation sites include or have included East Nunatuq, Arctic (Arctic River), Igloot, Tin Creek, West Tin Creek Hills, and West Tin Creek Flats. During the 2002 vote, Kawerak employee Julie Baltar pushed for quick site selection, believing it would facilitate funding. At that time the community nominated Tin Creek as the relocation site. Some people in the community do not feel that that nomination was legal or binding and disagreed with the site selection (based on interviews and personal communication). In 2010, Kate Kokeok reported that Tin Creek, West Tin Creek, and Nunatuq were the most community-supported sites.

All of these sites are challenged by the continuing climatic changes that progressively alter the landscape. Tin Creek, West Tin Creek, and Nunatuq all have ice and permafrost-rich soil, which is a poor foundation for infrastructure, particularly as the permafrost boundary moves north. Sites closer to the coast are at greater risk of continued erosion, but for the *Kigiqtaamiut* easy access to the coast is essential. Site selection is also sensitive because of Native allotments—parcels of land given to families who have rights to them through traditional use. Family camp areas are scattered throughout the mainland, and residents return to camps for seasonal subsistence harvests that mimic long-standing subsistence practices of going inland for fish, land mammals, and plant harvests and returning to the coast for sea mammals.

Residents of Shishmaref are worried about leaving the island at all to live inland. A primary concern is continued access to the sea for spring seal hunting. From the island, hunters can go directly from the village to the sea ice. On the mainland, residents would have to cross lagoon ice that could be "rotten," or unsafe to travel on, in order to get to the sea and the all-important spring seal hunt. Shishmaref residents say that they would return to the old tradition of camping during springtime—going to the island or some other point along the coast to camp on the ice to wait for the seals.

The seal oil and black meat from Shishmaref require coastal conditions to properly dry and render, according to sources. Preparing dry meat requires cooler temperatures and ocean breezes that protect the meat from insects and flies. As one woman told me, "even ten miles makes a big difference." If sites along the coast are considered unfeasible because of permafrost, and Shishmaref residents are pushed inland, then this all-important cultural tradition would be much more difficult, maybe impossible, to carry out.

As in 1901, choosing what is a viable site for an Iñupiaq sea mammal hunting community and what is a good site for state-sponsored infrastructure development may prioritize different objectives. Again we see practical considerations and ideologies clashing, and we might ask, what happens if all these criteria cannot be met? What makes a place "viable" for habitation depends on the value system one uses to assess a site. Obviously no one wants to build on a site that is subject to continued erosion, but site selection is not a straightforward process and must be handled carefully.

A DISCUSSION OF CONTEMPORARY RELOCATION

The history of infrastructure development in Shishmaref is a history of negotiation between action, inaction, and reaction by state and federal agencies and action, inaction, and reaction by local communities. Relocation discussions are a new chapter in these negotiations. The tools and strategies needed to adapt to erosion and flooding risks today are embedded in complex social networks, bureaucratic mandates and funding, how outsiders imagine and respond to ecological circumstances on an island in the middle of the Chukchi Sea, and how Shishmaref residents respond to the sites deemed feasible by engineering firms.

Ultimately, why Shishmaref residents live on an island exposed to flooding and erosion risks, and why they cannot easily relocate off of the island to mitigate the risks of these floods and prevent loss of life and property and diaspora is a function of history, climate change, ecology, colonialism, and cultural mandates. Vulnerability to disasters is the result of social systems interacting with ecologies—of risk entering into stratified social, political, and economic systems.

Everyone I've met who lives in or works personally with the village of Shishmaref acknowledges that something needs to be done to protect residents. All stakeholders agree, and this research makes clear, that there is institutional responsibility on the part of state and federal governments to give aid in securing that protection. Shishmaref residents have a right to access the modern institutions that came with colonialism—without having to sacrifice traditional landscape in order to do it. As Tony Weyiouanna testified simply to the US Senate: "We, as American people, deserve the attention and help of our fellow Americans."[161]

Chapter Six

The Tenacity of Home

Collocating or merging with other villages may be cheaper than relocation, but the risk is high that the village's lifestyle and culture will be lost. With these estimates, the Shishmaref Erosion and Relocation Coalition decided to continue with their relocation efforts.
—Climate Adaptation and Knowledge Exchange 2010[162]

This chapter is the one that mimics the anthropological tradition most directly and is the one I feel the least comfortable writing. What I'm calling here the *tenacity of home*, or the desire expressed by most residents to retain a village within traditional subsistence territory despite enormous challenges, is a deeply cultural, personal, and complicated subject. As an outsider, I neither fully understand nor can I explain the relationship between Shishmaref community members and their landscape. This story, however, would be incomplete without discussing the centrality of subsistence, practice, and relationships among people. The following chapter is an attempt at conveying to outsiders what I learned about the profound and untranslatable concept of "subsistence."

In interview data I collected as part of a cultural impact assessment for the Army Corps of Engineers from 2004 to 2006, 100 percent of fifty-four households interviewed responded that, if and when the community is relocated, the village should remain within traditional subsistence territory. Despite the varied opinions that men and women in Shishmaref have about relocation—which relocation site on the mainland and/or along the coast is preferable, how and when relocation should be carried out, who should lead the relocation effort, and how government representatives have handled relocation planning so far—there is the collective belief that Shishmaref should remain as a discrete community within the traditional *Tapqagmiut* area.

While worthwhile debates ensue about what rural Alaskan life will look like in the coming centuries, Shishmaref residents publicly make the argument that full removal from traditional subsistence territory would lead to cultural disintegration. Anthropologists have warned that it is dangerous for outsiders to see indigenous people as bound to geographical locations because it promotes a romanticized, alienating, and limited view of indigenous people. Iñupiat people who move to Anchorage or Seattle

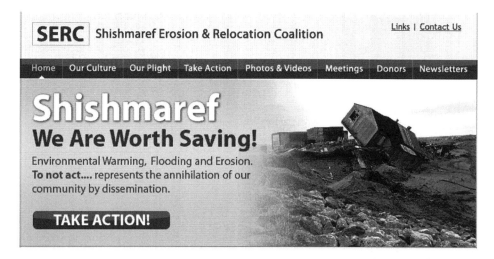

FIGURE 6.1   Erosion and relocation coalition banner

or Paris to become lawyers or doctors or engineers are still Iñupiat people, and Iñupiat cultural values are present and can be articulated in any occupation, any location, or any social context. Today, however, relocation outside of traditional territory means merging into a preexisting urban area and abandoning completely the territory that has kept the community together and bound to an ancestral past despite and during a tumultuous century. Place, in this situation, secures sacred social-ecological relationships. No one in Shishmaref that I've spoken to believes that closing the village completely—without establishing a new one in this particular, beautiful, and inhabited landscape—is a good idea. Most residents feel this would represent a significant challenge to maintaining cultural integrity. Figure 6.1 is the banner from the Shishmaref Erosion and Relocation website.[163] In the banner, Shishmaref residents present the argument that they are "worth saving."

The tenacity of home is complicated, but it is—at least in part—rooted in a subsistence tradition that is about food, practice, skills, and connection to family and place.

SUBSISTENCE PRACTICES IN SHISHMAREF

In 1985, Sobelman reported that Shishmaref residents obtained 75 to 80 percent of their total caloric intake through subsistence (defined as locally procured resources) proteins, fats, and vegetables.[164] In 1990, Conger and Magdanz reported that the average Shishmaref household took in 2,637 pounds of edible weight in subsistence foods during the year, or 663 pounds per person.[165]

> Marine mammals accounted for 69.4 percent of the total harvest, three times as much as any other resource category. The next largest component of the harvest

was land mammals (15.6 percent), followed by fish (6.4 percent), plants (3.4 percent), salmon (2.6 percent), birds (2.0 percent), and shellfish (0.7 percent).[166]

Subsistence practices in Shishmaref are integrated into all aspects of life. The annual cycle and daily activities of a household revolve around subsistence practices and the school day. To residents, the word *subsistence* refers to more than just the products derived from labor, but signifies a way of life and an orientation to and relationship with the landscape and with one another. Sharing customs, for instance, are different around subsistence activities as opposed to cash exchanges or other market activities. John Sinnok of Shishmaref says that subsistence constitutes a "we" world and is fundamental to an Iñupiaq way of life. Thomas Thorton writes:

> Every year tens of thousands of Alaska Natives harvest, process, distribute, and consume millions of pounds of wild animals, fish, and plants through an economy and way of life that has come to be termed "subsistence." Collectively, these varied subsistence activities constitute a way of being and relating to the world, and thus comprise an essential component of Alaska Native identities and cultures.[167]

The following is an excerpt from my field notes and reflections on days spent working with women friends to put away the *ugruk* from the spring hunt:

July 2008, field notes

July 2008 has been a rainy season.

Rachel postpones putting away the black meat that Dennis and other family members got from hunts in the spring. Her racks made of driftwood are covered. Stink hams (shoulder meat from bearded seals) are drying under tarps. Trashcans at the family's racks on the northwest, unprotected side of the island are full of seal blubber rendering into oil, butchered strips of thin, dried *ugruk* meat, and tightly wound intestines and stomachs. All the different parts of the *ugruk* are separated, and each has its own trashcan.

Finally, there is a break in the rain. Minnie (Rachel's sister in-law), Rachel and I go to the racks on Rachel's four-wheeler to put away the meat and to make buckets to be distributed to friends and family who live in Shishmaref and in other parts of Alaska.

Rachel sits the radio up on an old oilcan and flips a white five-gallon bucket over as a seat. She pulls the trashcans of blubber, dried black meat, stomachs and intestines, and the rendered oil out from where they were stored. Minnie sits next to Rachel. The two women divide the meat into smaller white buckets that are cleaned and prepared each year for this purpose.

The men have already hunted and begun to butcher. The women have finished butchering and divided the meat into these cans, so that the only thing left to do is to further divide the animal into buckets and submerge the dried meat with seal oil. A handful of stomachs and intestines go into a bucket, along with some thin strips of black meat, *panaaluk*, and thicker cuts of black meat, and then the bucket is filled with seal oil, opaque and yellow.

While we do this, the radio plays Casey Kasem's weekly top 40, and we all drink Orange Crush from cans. Rachel and Minnie don't talk much, but this is still a social event and there's an ease in spite of the hard work. The rain has finally let up and we can see the ocean and the mainland. It is beautiful here.

Like many times before, I'm at a loss to explain what this is like.

First, there is the repetition of the work itself—put in stomachs, intestines, *panaaluk*, seal oil, close bucket, repeat—routine and calming. Next it is overwhelming display of competency that Rachel and Minnie have while they work, a competency that comes, perhaps, not even from a lifetime, but from an accumulation of lifetimes in a family.

And there's the Orange Crush, the plastic five-gallon buckets, and radio pop music that show how nonresistant to change traditions in Shish actually are. Putting away black meat is ritual. It is a practice in cultural expression, but it is not static. It is not part of a living museum. Putting away black meat is what women do. It is life.

The *ugruk* we put away is the *ugruk* that Rachel's daughter Kate ate when she was pregnant and had morning sickness. This was the *ugruk* that Rachel's other daughter brought with her to Shaktoolik where she lives now with her husband and two sons. This is the *ugruk* that I would eat for the next week. This is the *ugruk* that Minnie would take home to her husband who does not hunt as much since his snow-machine accident. It's the *ugruk* that would be distributed to family and friends and would be used for special occasions and eaten when people didn't have the money to buy store food.

SUBSISTENCE AS RISK MANAGEMENT

Subsistence creates interdependence among cultural values, physical survival, and even the survival of the animals and landscape. It is a stretch to frame the profundity of this mutually constituting relationship among practice, people, and place as simply a strategy for risk management. It is more than that, but maintenance of the subsistence tradition is a pragmatic response to the uncertainties of life and daily risks people face. In Shishmaref, subsistence is even a risk-management strategy to protect the integrity of animal populations and ecological life. John Sinnok explains how the population of squirrels has gotten smaller since Shishmaref people have stopped hunting them:

When we were kids my grandmother used to have me do a lot of squirrel hunting for her, and she said, "if you guys quit hunting squirrels they'll disappear," and now that we don't eat as much squirrels as we used to, we don't hunt them, and now there's very little squirrels where there used to be a lot of squirrels when we were growing up because we hunted them for the skin and we ate the meat back then. Now we don't do that and you hardly see any. I'm not sure if they over graze or what's going on.[168]

In other interviews people reported that the land needed people or the animals would go away. In interviews some residents explained the ecological niche of *Kigiqtaamiut* people within their subsistence territory as being necessary for the survival and health of the land and animals.

Cultural tradition is maintained through subsistence, but this is inseparable from people's physical survival as well. In the following excerpt, for example, Raymond Weyiouanna ties subsistence to cultural survival but then quickly talks about the ocean as the place to get food, a material necessity. His use of the word "survive" at the end of this excerpt identifies, I think, the convergence of physical and cultural survival through subsistence.

Without subsistence, our lifestyle, our culture wouldn't be held together, I suppose. Because we depend on the sea for a lot of our food. The sea is like our supermarket—when the ocean is nice we gather what we can. When the ice is broken up whether it be the bearded seal, the walrus, and then after the ice goes we try to gather as much fish as we can from the sea, you know because it's calm, it's like the store is open when it's calm and like the supermarket is closed. Without that I don't think we'd be able to survive.[169]

Giddings noted that the Seward Peninsula has been a place of both continuity and change for millennia, and anthropologists write about flexibility to changing conditions as an important Iñupiat and Yupiit individual and cultural trait for dealing with the fluctuating environmental and social conditions in the Arctic.[170] Emblematic of this flexibility is the ease with which snow machines and other technology were incorporated into subsistence activities. I found, however, that hunting and subsistence were understood as a constant backdrop to persistent change. When research participants spoke about dynamic change in the Arctic, about adapting to changing scenarios, changing ecological conditions, and even moving to new locations, hunting and subsistence practices were expected to be flexible; but abandoning subsistence territory and not having access to the ocean *at all* was characterized as the breaking point of this flexibility. Fred Eningowuk outlines this belief explicitly by saying, "And then, you know Eskimos have always adapted to their location and their way of life. Eventually we would have to adapt to a new relocation site. Whether it be changing our subsistence way of life. The majority of us, you know, like me, would still need access to the ocean."[171]

Subsistence practices carry a strong cyclical quality. Again, Raymond Weyiouanna says:

> The *most important* [my emphasis] thing is to teach them [his children] is the value of the food that we provide them and the livelihood of having to teach them to learn how to get the animals and basically pass down what has been passed down to me from my parents, and that's what I'm looking forward to doing to my children.[172]

The importance of passing on subsistence practices to the next generation, and the burden of responsibility to learn subsistence skills were present in many exchanges I had in Shishmaref. In another interview, Esther Iyatunguk said the following:

> And my aunt, she would always give my mom and dad a little bucket. She shared. I don't know if you met her, Sharon Nayokpuk, she was like an older sister. I noticed last year she was getting tired a lot when we started cutting, and I helped her. You know I helped her cut last year, and I'm going to help her this year because she helped our family a lot. You know it's my time to help take care. It's just our time to step up, you know?[173]

Esther used this phrase, "time to step up," frequently in reference to subsistence activities. Esther has a job with the school, has five children, and takes online classes through University of Alaska Fairbanks Northwest campus. She never said she was "stepping up" when she talked about her work or education. Instead she used this idea of "stepping up" to talk about her brother learning how to seal hunt, her learning to making *kuspuks* after her Gram passed away, and her helping with the cutting of seal and *ugruk* meat.

By passing these traditions on, subsistence hunting becomes the constant in a dynamic world, even if the form of hunting changes. The strong generational component of subsistence hunting is present in Shishmaref, and it is incorporated into new social media and new social spaces. On October 20, 2012, Kate Kokeok posted a photograph of her son's first seal catch on Facebook. One hundred and forty-nine people "liked" it. Even as many Iñupiat people prize flexibility and incorporate modern technology, infrastructure, and ideologies for their own use and expression, the constancy of subsistence can be seen as a rational strategy for mitigating the fluctuations in economy, politics, and social life that have marked a century of radical change.

Throughout the Seward Peninsula, people discuss the day when villages will have to be completely self-sustaining. This narrative never surfaced in interviews, but it came up in personal and intimate settings. There is a strong belief that the white settler population will one day leave, as well as the airplanes, Department of Transportation money, subsidized electricity and so on. This day, residents understand, will be catastrophic—as the dismantling of public infrastructure would be for any US community. Acknowledging and being mentally prepared for this is part of being prepared for a dynamic and changing world in which social and ecological circumstances are not entirely predictable. Being at home—or at least having *someone* at home, in a place where subsistence can be carried out—is a measure of food security and a measure of security for the future of the family group in the face of catastrophe.

Considering that subsistence practices have been a reliable (and preferred) food source for thousands of years and that supermarkets have a much shorter history this is a rational strategy of risk management and economic diversification. In Shishmaref, starvation times still exist in living memory, and supermarkets are still not a primary food supplier to most Shishmaref residents. Given these circumstances, being divorced from the ability to provide food for your family through subsistence is dangerous; being away from the ocean, as Eningowuk claims, is not an option.

INTERDEPENDENCE AND PLACE IN RURAL ALASKA: WHERE I LIVE IS NOT A CHOICE

Nusugruk Rainey Hopson is an Iñupiaq freelance writer, blogger, and artist from Anaktuvuk Pass. Recently she engaged in a public discussion and debate with someone [LR] "from the outside" (the forty-eight continental US states) about living in rural Alaska and about whether or not the expense of heating a house in the rural Arctic, where heating oil is expensive, was "worth" it:

> LR: I'm a "people of the lower 48" . . . Rainey has told me quite a bit over the past few years about the harsh weather and the high cost of items. *It's a personal choice to stay living there* [my emphasis], as it is a personal choice for anyone to live in whatever state they live in. I hear Alaska has some pretty scenery, but I can hardly afford to heat my house through 10 to 20 degree winters. Though I imagine surely in −50 weather a fireplace doesn't cut it.
> NRH: It's not a personal choice actually, which is hard to define to people because it's such a culturally defined decision. In our culture, how we are raised, what we see every day, ties us to this land. It's the opposite of the "independence and separation" type of culture found in most places in the lower 48. Here it's central, the connection with land and animal and family. I think when your family lives in the same spot for over 10,000 years, the culture surrounding that heritage makes your "personal choice" to live here or not null and void.

In the above passage, Hopson is frustrated. This tone is apparent in interviews in Shishmaref surrounding issues of relocation and subsistence as well. Some Shishmaref residents expressed similar frustration when asked about *why* they needed to stay on traditional land and conduct subsistence practices. Annie Kokeok said, "it's just our way of life, the subsistence way of life." The frustration comes from, as Hopson says, being "hard to define to people because it's such a culturally defined decision."

The value of a subsistence lifestyle is predicated on an Iñupiaq orientation toward the world and differs quite profoundly from non-Iñupiat worldviews. Therefore, explaining the importance of maintaining small, traditional, rural villages and landscapes can be difficult in cross-cultural settings. It is apparent that for many people with an Iñupiaq cultural orientation, subsistence is necessary for maintaining personal relationships, cultural continuity, and physical security—whether or not this makes sense to outsiders.

During an interview with Brice Eningowuk, he comments that Shishmaref residents "can't get away" from hunting or subsistence practices, regardless of the uncertainty about the future or about various scenarios of relocation following a disaster. "We're [*Kigiqtaamiut* people] going to be hunting; we're going to be doing subsistence no matter what, I think. That's one of those things I feel we can't get away from."[174]

Consider this statement from Raymond Weyiouanna: "Without subsistence our lifestyle, our culture wouldn't be held together, I suppose." Subsistence in these examples is not something people choose but rather a set of relationships people have with their communities and a responsibility people have to one another and the places from which they come.

I DON'T WANT TO MOVE, EVEN IF I MOVE

In interviews conducted from 2004 to 2006, residents reported unanimously that they did not want to move to Nome or Kotzebue.[175] When I returned to Shishmaref in the late 2000s, I learned that some of these same people who were interviewed had moved to Nome. Similarly, some of the most passionate and dedicated Iñupiat people I know do not currently live in the villages in which they were born. What is the cause for this seeming discrepancy in speech, cultural mandate, and action?

In the 2004 to 2006 interview script, one of the first questions asked of participants was whether or not they wanted to move to Nome or Kotzebue. The interview script assumed the individual as the basic unit of analysis—but this may not have been the unit of analysis perceived by all interviewees, particularly if they responded "no," only to move away shortly thereafter.

In a classic essay on the nature of anthropological understanding, Clifford Geertz argues against assuming the primacy of the independent (or individual) self in our cross-cultural endeavors:

> The Western conception of the person as a bounded, unique, more or less integrated motivational and cognitive universe; a dynamic center of awareness, emotion, judgment, and action organized into a distinctive whole and set contrastively both against other such wholes and against a social and natural background is, however incorrigible it may seem to us, a rather peculiar idea within the context of the world's cultures. Rather than attempt to place the experience of others within the framework of such a conception, which is what the extolled "empathy" in fact usually comes down to, we must, if we are to achieve understanding, set

that conception aside and view their experiences within the framework of their own idea of what selfhood is.[176]

The emerging field of cross-cultural psychology continues in this vein, pointing out that many non-European-based cultures construct the self and the agency of the self with a focus either on independent or on interdependent relationships.[177] While "every individual self carries elements of independence and interdependence,"[178] the degrees to which the former or latter provide underlying structures for organizing social behavior vary between cultures. In other words, there are both practical and cognitive distinctions between cultures in which independence is the basis for social life and cultures in which interdependence is a foundation for social life.

Applying this framework to the interview script and the answers I received to my initial interview question, I might conclude that the "I" that participates in the community of Shishmaref does not want to be relocated to Nome or Kotzebue, regardless of whether or not the individual "I" does actually move or desires to move to Nome or Kotzebue. Shishmaref residents want Shishmaref to exist even if they live somewhere else. The individual is in part constituted by the existence of Shishmaref as a social unit; thus agency is interdependent, and the interdependent self does not want to relocate.

The analytic tool of independence and interdependence used by cross-cultural psychologists is rudimentary and subsumes a wide variance of cultural expression. However, it also sheds light on Hopson's explanation that living in the Arctic landscape is "not a choice," by which she may mean it is not an independent choice but rather an interdependent one. It also may explain Iyatunguk's statement that it was "time to step up." Under certain conditions it becomes a responsibility and obligation of the individual to participate in the social structure of the village—to align personal agency with participation of the group. So when asked, "Do you want to move to Nome or Kotzebue?" the answer is "no," regardless of whether a particular individual moves to Nome, Anchorage, or California.

EXTENDING INTERDEPENDENCE TO THE LANDSCAPE

The interdependence of personhood in the cultural psychology literature is exclusively a set of relationships and orientations among people, but Ingold asks,

> What makes a relationship social, and are such relationships confined to human beings? Why should it be supposed that we encounter the nonhuman components of our environment—animals, plants, inanimate objects—in their sheer materiality? What do we mean by saying that our relations with these components are material relations? Or to put the question in its even stronger, converse form, what does it mean to say that these relations are not social?[179]

In other words, Ingold asks, can our relationships with the environment be social? Can they be interdependent? Trying to understand why tenacity and an abiding dedication to the landscape is such a prominent feature of Shishmaref social life, articulated explicitly during this period of high risk, necessitates investigating fundamental assumptions about people's relationships with their environment. Northern indigenous scholars often report that many ethnic groups attribute agency to nonhuman things, including animals, landscape, and weather.[180] Animism in this context is not the imbuing of spirit into animate things but rather is a conception of the material world as being inseparable from what we may term "spirit":

> Animacy, then, is not a property of persons imaginatively projected onto the things with which they perceive themselves to be surrounded. Rather—and this is my second point—it is the dynamic, transformative potential of the entire field of relations within which beings of all kinds, more or less person-like or thing-like, continually and reciprocally bring one another into existence. The animacy of the lifeworld, in short, is not the result of an infusion of spirit into substance, or of agency into materiality, but is rather ontologically prior to their differentiation.[181]

Describing engagement with the landscape and with animals as engagement with an animated and spirit-infused life-world seems to be an accurate way to describe the relationships between people and the environment in Shishmaref. Josh Wisniewski writes about *anjzugaksrat iniqtigutait*, the set of rules and laws used in Shishmaref that govern right action in the world and have particular salience for hunting luck and success. Under *anjzugaksrat iniqtigutait*, or "Eskimo law," *sila*, often translated as weather, is actually conceived of as the "environment, the organization of the world, consciousness, and weather without implying a differentiation between these conditions of the world."[182] *Sila* under this translation is animate, as understood by Ingold in the passage above. The world was imbued with agency prior to differentiation into humans, animals, and landscape. If relationships among sentient beings—including humans, animals, and landscape—are what interdependently construct agency and personhood, then the landscape and relationships with the landscape literally, not figuratively, are definitive of *Kigiqtaamiut* people and culture. If relations are what "bring people into existence," according to Ingold, then we can understand the tie between *Kigiqtaamiut* people and landscape as a mutually constitutive relationship.

Translating these cultural imperatives into bureaucratic frameworks is extremely difficult. Subsistence occupies an uncomfortable terrain in agency reports concerning the relocation of Alaska Native villages that are exposed to erosion and flooding. Subsistence is not quite defined as an economic imperative—though it competes with and subsidizes market labor and is in some ways a function of the economy—and it is not solely symbolic or recreational either.

A report from the IAWG contrasts "jobs" with subsistence opportunities: "BLM firefighting, construction work, and other seasonal jobs often conflict with subsistence *opportunities* [my emphasis]."[183] In this case, subsistence is distinct from economies but still exists within a worldview in which an individual may take advantage of "opportunities." In the next passage from the IAWG, the report identifies Alaska Native peoples as *interested* in culture and tradition, which includes subsistence. "Remote Alaska villages typically are largely native, and have a significant *interest* [my emphasis] in culture and tradition."[184] Shishmaref residents, as demonstrated by the preceding interviews, make much stronger statements about the importance of remaining on traditional land and having access to subsistence territory. The banner for the Shishmaref Erosion and Relocation Coalition strongly indicates that the connection among people, animals, and land is more vital than an "opportunity" or an "interest." The banner emphasizes that "We" are worth saving—encompassing the entirety of this triad in Shishmaref.

A CONCLUSION: TAKING INDIGENOUS PEOPLE SERIOUSLY

In his article *The Gift of the Animal*, Paul Nadasdy writes that there would be a radical shift in anthropological theory if anthropologists accepted as an actual, not merely metaphorical, truth that humans and animals could have social relationships with one another.

> In short, we must acknowledge that they are not just cultural constructions and accept instead the possibility that they may be actually (as well as metaphorically) valid. For the most part, however, we have refused to do this. In this article I take seriously the possibility that northern hunters' conceptions of animals and human-animal relations might embody literal as well as metaphorical truths.[185]

This idea of taking people seriously—literally and not metaphorically—applies in the Shishmaref case study as well. Even without an emic understanding of what subsistence means and how relationships within *Kigiqtaamiut* culture are constructed, it is important to believe that residents *say* removal from this place and into a more urban environment is not an option. Agency planning should begin with the first imperative of Shishmaref residents themselves: "We must have access to the ocean." While there are no current plans to move *Kigiqtaamiut* people off of traditional land, forced relocation and the consolidation of indigenous groups throughout the circumpolar North has been a consistent trend for the last one hundred years.[186]

Shishmaref residents know this history—many people have lived through it—and, following, residents express significant fears that diaspora, dispersal, and integration into a larger community will be the outcome for their own community. These threats have become particularly looming because of the geographically widespread flooding and erosion risks being experienced by Alaska Native communities today. If we accept that Shishmaref residents must (as they say) remain on traditional land, then the answers to complicated logistical and economic questions should stem from this local imperative.

A case for Shishmaref residents remaining on subsistence territory should not be framed in terms of a cost/benefit analysis—predicated on percentages of subsistence foods that make up household caloric intake or the amount of transfer payments versus household financial independence. Rather, solutions to the expense of service delivery and risk mitigation can be analyzed by first considering whether or not Alaska Native peoples have a right to traditional territory, a right to rurality, and a right to subsistence. Solving complex problems can take place under this rubric.

This chapter addresses the tenacity of home—*why* Shishmaref residents say they need to stay near the ocean. Much more important than this is the call to take Shishmaref residents seriously, whether or not the theories presented here are accurate or have explanatory power. Unanimously, Shishmaref residents who were interviewed said that they did not want to merge the village with a larger community but wanted instead for Shishmaref to remain a discrete village within traditional territory. Whether or not this cultural imperative is understood through theories of interdependence and animism is secondary to the fact that this imperative is explicit and decisive. Shishmaref residents' demonstrated resolve is enough to make remaining on subsistence territory a bureaucratic mandate.

Chapter Seven

The Ethics of Climate Change

One might be tempted to say that these resettlement projects fail because of the shortage of resources and skills that plague developing countries. Let us not forget how resource-rich countries such as Canada and the United States have disrupted their marginalised "native" populations through resettlement projects. If North America can't make it work, you might ask, what hope do third world countries have?

—Chris De Wet 2001[187]

It is difficult for those of us professionally involved in Indian policy to comprehend the level of unimportance that Indian law and policy has occupied on our scale of national priorities. . . . It is particularly ironic that despite a generally low level of national attention, a great many people not only claim familiarity with but readily volunteer answers to questions concerning Indian affairs.

—Rennard Strickland 1979[188]

That under-the-ice netting has been going on for years and years. Not nothing new. We have a system, a way of doing it.

—Interview with Clifford Weyiouanna, July 21, 2008.[189]

A SUMMARY OF THE SHISHMAREF CASE STUDY

This research is an ethnography of climate change—a sketch of the complex factors historically and contemporarily that create vulnerability and low adaptive capacity in Shishmaref to erosion and flooding that result from a warming world. Merging historical data (including data from the oral and written records) and contemporary experiences of vulnerability (through ethnography, interview, and survey) is a way to capture in the present the individual, community, and global movements of history, society, and environment as they play out in one particular location. This is grounded case study research. A research project that engages this scope and perspective on climate change is a unique contribution to the field of anthropology, and in the end this research is a holistic study of a moment in time when *Kigiqtaamiut* people and other Shishmaref residents are waiting for what will happen next—the study, the storm, the organized relocation, or the disaster.

The Seward Peninsula has been inhabited for at least four thousand years by dynamic and flexible cultures that have adapted to changing ecological conditions, adopted new technologies, moved around, and moved on. Today in Shishmaref, continued erosion and flooding and the ineffective long-term viability of shoreline stabilization make migration off the island and resettlement elsewhere the only reasonable solution for long-term safety.

Migration itself is not a maladaptive strategy for coping with ecological shifts—conversely, for millennia migration has been a successful strategy for accommodating ecological shift.[190] However, the last one hundred years of displacement and resettlement, particularly of indigenous and marginalized groups, have mostly been a failed experiment in government-driven, social engineering that resulted in the further impoverishment and social disarticulation of moving populations.[191] In light of these resettlement failures, it is critical to understand the outcomes residents are trying to avoid. Vulnerability in Shishmaref is not exposure to rising waters and falling bluffs but is rather that, subsequent to rising water and falling bluffs, Shishmaref residents will experience negative outcomes. In the event of a large storm, Shishmaref residents are likely to be threatened with loss of life and loss of property. In the long term these risks must be mitigated through relocation.

Diaspora and dispersal out of traditional subsistence territory is the single greatest fear of residents I interviewed in Shishmaref. *Kigiqtaamiut* people themselves see removal from subsistence territory as a mechanism of cultural disintegration and the possible disintegration of the landscape as well. There is a complex relationship among people, society, and landscape in Shishmaref. Regardless of the academic understanding of this relationship, it is unequivocal that residents see the dispersal of Shishmaref residents as *increasing risk* to themselves and their cultural heritage. This position should be taken seriously.

Recurrent throughout American Indian and Alaska Native policy in the United States is the imposition of outsider ideologies through infrastructure and institutions, which either failed to create any sustained improvement in the lives of American Indians and Alaska Natives or made the quality of life worse. As Rennard Strickland writes, "a recurring historical fact is that Indian policy makers have believed, or acted as if they believed, that Indians did not know what was good Indian policy.[192] Today, Shishmaref residents believe that reconstructing the village on the mainland and/or on a nearby coastal island is the best solution in order to mitigate risk and remain on traditional subsistence territory. Shishmaref residents know what is good Shishmaref policy.

RAPID ECOLOGICAL AND SOCIAL CHANGE

The decision Shishmaref residents are making to stay close to home occurs against a backdrop of dramatically changing social and ecological conditions. Social changes over the last one hundred years significantly contribute to flooding risks in the present. In the past, mobile infrastructure made high mobility and adaptation possible for

Tapqagmiut people. Today, Shishmaref residents are less mobile due in part to modern infrastructure that is incorporated into the daily lives and basic service needs of community members.

The institutions and infrastructure that have been integrated into the daily lives of the *Kigiqtaamiut* (as with the entire citizenry of the United States) were expensive to build. This was true in 1901 and remains true today. Sheldon Jackson petitioned outsiders for additional funding for school projects on the Seward Peninsula because federal funding for infrastructure was inadequate even at the turn of the century. Original infrastructure investment in Shishmaref was justified by the ideological belief that modernization, Christianization, and civilization would benefit Alaska Native tribes and was part of a global colonizing project that occurred throughout much of the world. Subsequent infrastructure, development, and service delivery projects in Shishmaref (such as an airport, electricity services, and a barge landing) were funded or subsidized by the government as standard practice for rural service delivery in the United States.[193]

Delivering the services of high modernity is expensive in rural Alaska. At first glance it may seem like rural Alaska villages are not economically viable, but these institutional expenses are a relatively minor colonial trade-off. From an economic perspective, colonization is cost effective for the colonizers—and this is overwhelmingly true for resource-rich Alaska.[194] The financial burden of flooding now is a cost incurred by the colonial model and, as such, places the burden of responsibility on the same institutions that pushed for infrastructural development in the first place—namely state and federal governments.

VILLAGE VIABILITY

An underlying issue for villages that need to relocate because of climate-change-related erosion and flooding is whether or not Alaska Native rural villages are viable in the twenty-first century. As the number of villages exposed to erosion and flooding increase, and as cost estimates for relocating a single village top $200 million, it often seems that the unspoken question is why these villages, some with populations as small as eighty people, exist in the first place. Urbanization into larger economic hubs can seem like a rational plan for small villages without running water that face increased risk.

The urbanization of Native American peoples is a consistent trend in federal policy. The termination and relocation policies following World War II were successful in moving large numbers of American Indians out of reservations and into urban areas. This was explicitly a federal attempt to assimilate and increase employment among American Indians. Eventually these policies were "widely attacked, especially by American Indian advocate groups,"[195] and most policies were halted or reversed by 1975. In the Arctic, consolidation of Alaska Native and Siberian Native settlements occurred through both Soviet and American government projects.[196]

A real question is not whether climate change and flooding risks will be *a* catalyst forcing Alaska Native peoples to urbanize or to relocate out of traditional land but whether climate change and flooding risks will be the *next* catalyst for forcing Alaska Native people to urbanize and relocate out of traditional land. With this historical grounding, it is exceedingly clear that Alaska Native villages and settlements have been fighting against disintegration and for recognition as "viable" entities since the colonial project began in earnest.

The waves of relocation in American Indian and Alaska Native communities are rife with issues of social justice and demonstrate the continued marginality of minority and rural populations. That these communities have to justify their existence in the face of climate change—and that these communities find themselves at greater risk than other communities—demonstrates that vulnerability is the product of *systems of inequity*, not a characteristic inherent in a single community.

WHY THE PUBLIC SHOULD CARE ABOUT SHISHMAREF

This research set out to address the issue of climate change and related flooding and erosion in Shishmaref, but equally important to this primary focus are the inevitable questions that follow. Namely, what can be done about vulnerability and risk in Shishmaref, and why should anyone outside of Shishmaref care? The answers to what creates vulnerability and what can be done about it are inextricably linked. By understanding the social, ecological, and infrastructural building blocks that create vulnerable communities, we can understand how best to build resiliency and adaptive capacity and lower vulnerability in at-risk communities. The answer to why the general public should care is both more challenging and more critical. In this case, the limits of scientific inquiry intersect with the beginnings of an ethical dilemma that will not likely be answered satisfactorily with research and ever more bits of data and information.[197]

Climate change itself presents a monumental ethical dilemma to global residents. From what we know about disaster and vulnerability, we can predict that marginalized and already vulnerable populations are more likely to experience negative outcomes of climate change than their resilient counterparts—research so far has found this to be overwhelmingly true.[198] These communities are the least likely to have produced large amounts of greenhouse gas emissions that cause anthropogenic warming. In cases like Shishmaref, the burdens of moving are linked to changing ecological conditions, and this raises questions about how burdens of anthropogenic warming are and will be distributed.

The ecological risks in Shishmaref also raise profound ethical questions about Native American rights to traditional homelands. In Alaska, Alaska Native peoples often hold title to their land through the corporation system that was developed through the Alaska Native Claims Settlement Act of 1971 (ANCSA). As landscapes change and places become uninhabitable, a question arises: Do Alaska Native peoples have the right to real and realistic access to these land claims and the right to remain on traditional territory?

Based on the information and analysis presented in this case study, I believe the answer is yes. The history of Shishmaref demonstrates the rapid social changes that have occurred in the last one hundred years. From development and colonization to boarding schools, Alaska Native peoples have been outstandingly flexible to rapid social shift, but this flexibility has limits. Both formally and informally, Shishmaref residents make these limits explicit—saying that removal from traditional land is equivalent to cultural disintegration. Failure to take seriously the threat of cultural disintegration in Shishmaref is unethical. Real and realistic access to traditional territory—a requisite condition for Shishmaref residents to maintain cultural identity—should be an inherent and acknowledged right for the *Kigiqtaamiut*. Whatever the future for Shishmaref peoples, it should happen under direct mandate of local leaders and within appropriate cultural frameworks.

SUGGESTIONS MOVING FORWARD

In general, ethnographic studies like this one help elucidate the outcomes of political and social choices so that we can act on the ethical dilemmas surrounding climate change and disaster with greater awareness and understanding. In-depth ethnographies of vulnerability and disaster are still rare in anthropology. Future research should be directed at filling this gap. Only with robust comparative case studies will we be able to conduct meta-analysis on disasters and vulnerability. In Alaska, the next research agenda may be an investigation of what creates resiliency in Shishmaref—a methodological project designed around what to foster and how to build capacity in communities that need to relocate instead of those social variables that help create risk. For now, I offer six suggestions for moving forward.

1. Climate Change Demands New Disaster Response Protocol

The governance structure for disaster response in the United States through Homeland Security and FEMA is ill equipped to handle disasters that occur as a result of permanently changing ecological conditions. Because disaster response to date is based on protocol that emphasizes rebuilding in place, this does not allow for flexibility when ecologies and landscapes enter new states of normal, as they will with the continued onset of anthropogenic climate change.

Further, the only FEMA-sponsored hazard mitigation strategy that incorporates migration and relocation in order to reduce risk is only applicable at the individual household level and does not accommodate community-wide migrations. This is insufficient when entire communities need to be relocated, and it is insufficient in communities (such as indigenous communities) where there are multiple reasons (social, cultural and economic) to stay or relocate as a group in order to avoid dispersal.

2. Create a Central Agency for Relocation Planning

This research found that the turnover rate was extremely high among agencies and agency workers who were tasked with the relocation of Alaska Native communities impacted by increased erosion and flooding risks. This high turnover rate raises serious questions about institutional memory at the state and federal level. Shishmaref residents have experienced multiple iterations of "government help" and have become fatigued by inexperienced workers. This leads to miscommunication and inefficiency.

Protocol for relocation of communities faced with habitual flooding should be developed at a state or federal level. A central agency or policy should be developed to avoid redundancy, improve efficiency, and give structure to ad hoc relocation efforts happening today in multiple communities. The ACCIMP could possibly fill this role, but that has yet to be determined. This research calls for a central agency and policy protocol for village relocations; *however*, I note that some relocation leaders in Kivalina resist this centralization, fearing this would result in a lack of local control over relocation decisions. In Shishmaref as well, community members overwhelmingly insist that final decision-making power rest within the community. This must remain true, as I note below.

3. Work Closely with Local Institutions

> *A clear finding of the literature on resettlement has been that too often the process has been a 'top-down' one in which the involvement of those being displaced has been limited.*
>
> —Graeme Hugo 2011[199]

Over the last century, top-down planning resulted in forced displacement, which has led to long-term social problems for migrating communities. This research found that relocation planning at the state level often did not coordinate with planning at the local level. Schematically, these situations mean that at different scales of intervention (local, state, federal, international), institutions are working against one another. Local participation becomes compromised when decisions are made in Anchorage or Juneau and not in conjunction with local leaders.

Shishmaref residents express a strong desire for self-determination with regard to relocation planning. This is a highly valued priority. Any successful and efficient relocation planning requires joint efforts from multiple institutional levels but specifically requires *meaningful* local participation. To ensure meaningful local participation, there should be a priority on in-person communication among local, state, and federal leaders. Meetings and information-sharing events should take slow technology, non-English speakers, and unique cultural institutions (such as the elders' council) into consideration.

4. Develop Mechanisms to Encourage Personal Cross-Agency Relationships

This research suggests that the most successful way to accomplish cross-agency communication and multiscale efficiency and understanding is to encourage long-term personal relationships between agency workers and local leaders. These long-term personal relationships are the best mechanism for fostering progress by encouraging efficient, culturally appropriate communications, avoiding redundant research and planning, lengthening multiscale institutional memory, and finding creative solutions for moving forward. Long-term personal relationships may bridge the gap between the realities of village life and the lives of agency workers in Anchorage and Juneau and could satisfy Shishmaref residents' requirement that bureaucrats "see for themselves" the risks that rural communities face.

5. Outline Risks and Outcomes in Explicit Terms (While Recognizing Culturally Divergent Value Systems)

In Shishmaref, relocating residents away from flooding risks is not a sufficient adaptation plan. Instead, Shishmaref residents need to work with state and federal agencies to identify and create viable, culturally sensitive futures. Keeping discourses about relocation explicit in terms of what to avoid and what to accomplish is crucial in creating real, long-term adaptations to changing conditions. In the case of Shishmaref, this means selecting a relocation site on the mainland that offers practical access to the coast.

All sites are not equal. What constitutes an appropriate site to promote continued subsistence hunting and an appropriate site from an engineering perspective may differ. In order to avoid the mistakes of the past (such as ignoring local knowledge), communication among agencies and local residents must be sensitive to differences in value systems and site requirements. Compromises that may have to be made regarding an appropriate site will be difficult; communication among agency workers and local residents must therefore be as meaningful and precise as possible. Explicitly outlining risks, outcomes, and goals is vital in these communications.

6. Acknowledge an Alaska Native Right to Traditional Subsistence Territory—and Fund the Protection of This Right

Flooding and erosion happening today is directly related to the colonial project and, as such, is a cost incurred by colonial institutions and state and federal governments. This culpability, paired with the expressed desire by residents to remain on traditional land, means that the state should acknowledge an Alaska Native right to realistic and sustained access to traditional subsistence territory and property claimed through the

ANCSA process. Formal acknowledgement of a Native right to access traditional land could act as a starting point in planning the relocation process.

A FINAL WORD

In December 2014, I flew with Tony Weyiouanna and his wife Fannie to a conference in Ottawa, Canada, to discuss what was going on in Shishmaref. Again, as so many times before, people were enthralled by the story Tony told about his community and the ecological changes happening there. Again, Tony told a balanced story of the community that was not hyperbolistic, that included multiple perspectives, and that pointed to mechanisms that could support and organize relocation. Afterward, discussion among some audience members focused, as it always does, on the cost of relocation. This is a narrow perspective. As Peter Evans, the great anthropologist, said to a room of skeptics after a talk I gave on Shishmaref early on in this research, "it's not that much money."

It's not that much money. Compared to what relocation would save, compared to what colonization cost, and compared to the histories of genocide and resource extraction in Alaska and the culpability of governments for creating risk in the first place, relocation is absolutely affordable. What the United States and the state of Alaska can do today to keep climate change from grossly overburdening a population that did almost nothing to cause it is to fund the relocation of critical infrastructure for six hundred people, ten miles across a lagoon, to a safe place of their choosing, close to the ocean. This is not philanthropy—this is one meager step toward justice.

Notes

1. Interview conducted by Elizabeth Marino with Herbert Nayokpuk in Shishmaref, Alaska, summer 2005.
2. The usage, as reflected throughout this book, is Iñupiaq (singular) and Iñupiat (plural). The adjective form must agree in number with the noun it modifies, though this is often not an issue in the Iñupiat language because adjectives and nouns become the same word (a subject for a different book!).
3. ACIA, *Arctic Climate Impact Assessment* (Cambridge: Cambridge University Press, 2005), 992.
4. Henry P. Huntington, "Native Observations Capture Impacts of Sea Ice Changes," *Witness the Arctic* 8, no. 1 (2000): 1–2.
5. Larry Hinzman, Neil Bettez, W. Robert Bolton, F. Stuart Chapin, et al., "Evidence and Implications of Recent Climate Change in Northern Alaska and Other Arctic Regions," *Climatic Change* 72 (2005): 251–98.
6. ACIA, *Arctic Climate Impact Assessment*, 997.
7. United States General Accounting Office (USGAO) Alaska Native Villages: Most Are Affected by Flooding and Erosion, but Few Qualify for Federal Assistance. (2003) Report to Congressional Committees, *www.gao.gov/new.items/d04142.pdf*, (08.25.2011).
8. Nicole E. M. Kinsman, Meagan R. DeRaps, and Jacquelyn R. Smith, *Preliminary Evaluation of Coastal Geomorphology and Geohazards on "Kigiqtam Iglua," an Island Northeast of Shishmaref, Alaska*. Alaska Department of Natural Resources, Division of Geological & Geophysical Surveys, 2013.
9. For reviews of this literature, see Étienne Piguet, Antoine Pécoud, and Paul De Guchteneire, "Introduction: Migration and Climate Change," in *Migration and Climate Change*, ed. Étienne Piguet, Antoine Pécoud, and Paul De Guchteneire (Cambridge: Cambridge University Press, 2011), 1–33; and Lezlie Morinière, "Tracing the Footprint of 'Environmental Migrants' Through 50 Years of Literature," in *Linking Environmental Change, Migration and Social Vulnerability*, ed. Anthony Oliver-Smith and Xiaomeng Shen (Bonn: UNU-EHS, 2009), 22–31, *www.ehs.unu.edu/file/get/4019*, accessed February 27, 2015.
10. Piguet et al., "Introduction: Migration and Climate Change," 3.
11. Essam El-Hinnawi, *Environmental Refugees* (Nairobi: United Nations Environmental Programme, 1985), 4.
12. Intergovernmental Panel on Climate Change (IPCC), "Policymakers' Summary," in *Climate Change: The IPCC Impacts Assessment*, ed. W. J. McG. Tegart, G. W. Sheldon, and D. C. Griffith (Canberra: Australian Government Publishing Service, 1990).

13. Norman Myers, "Environmental Refugees in a Globally Warmed World," *Bioscience* 43 (1993): 752–61.

14. International Organization on Migration (IOM), *Migration and Climate Change*, Migration Research Series No. 31 (Geneva: IOM, 2008); IOM, "Climate Change, Environmental Degradation and Migration," *www.iom.int/jahia/webdav/shared/shared/mainsite/microsites/IDM/workshops/climate-change-2011/background_paper.pdf*, accessed October 9, 2011.

15. United Nations, *Population, Environment and Development* (New York: United Nations Department of Economic and Social Affairs, 2005).

16. Myers, "Environmental Refugees," 752–61.

17. Norman Myers and Jennifer Kent, *Environmental Exodus: An Emerging Crisis in a Global Arena* (Washington, DC: The Climate Institute, 1995).

18. IOM, *Migration and Climate Change*.

19. See http://en.wikipedia.org/wiki/Environmental_migrant, accessed November 12, 2012.

20. Simon Batterbury and Andrew Warren, "The African Sahel 25 Years After the Great Drought: Assessing Progress and Moving Towards New Agendas and Approaches," *Global Environmental Change* 11 (2001): 1–8.

21. This information comes from multiple interviews with local residents as well as the US General Accounting Office (GAO), *Alaska Native Villages: Most Are Affected by Flooding and Erosion, but Few Qualify for Federal Assistance*, accessed August 25, 2011, *www.gao.gov/new.items/d04142.pdf*; GAO, *Alaska Native Villages: Limited Progress Has Been Made on Relocating Villages Threatened by Flooding and Erosion*, accessed August 25, 2011, *www.gao.gov/new.items/d09551.pdf*; US Army Corps of Engineers (USACE), *Alaska Village Erosion Technical Assistance Program: An Examination of Erosion Issues in the Communities of Bethel, Dillingham, Kaktovik, Kivalina, Newtok, Shishmaref, and Unalakleet* (Anchorage, AK: USACE, 2006).

22. GAO, *Alaska Native Villages: Limited Progress*, 10.

23. G. Hufford and J. Partain, "Climate Change and Short-term Forecasting for Alaskan Northern Coasts," in *Proceedings of the American Meteorological Society Annual Meeting*, January 2005, San Diego, CA.

24. Orson Smith and George Levasseur, "Impacts of Climate Change on Transportation in Alaska," in *The Potential Impacts of Climate Change on Transportation* (Washington, DC: US Department of Transportation, 2002).

25. S. M. Solomon, D. L. Forbes, and R. B. Kierstead, "Coastal Impacts of Change: Beaufort Sea Erosion Study," Canadian Climate Center Report No. 94-2 (Toronto: Atmospheric Environmental Service, 1994); and James Syvitski, "Projecting Arctic Coastal Change," in D. L. Forbes, *State of the Arctic Coast 2010—Scientific Review and Outlook* (Potsdam, Germany: International Arctic Science Committee, 2010).

26. Fouad N. Ibrahim, "Causes of the Famine Among the Rural Populations of the Sahelian Zone of the Sudan," *GEO Journal* 17, no. 1 (1988): 133–41.

27. Ernest Burch, *Eskimo Kinsman: Changing Family Relationships in Northwest Alaska* (Saint Paul, MN: West, 1975); Ernest Burch, *The Iñupiaq Eskimo Nations of Northwest Alaska* (Fairbanks: University of Alaska Press, 1998); and Kathryn Koutsky, *Early Days on Norton Sound and Bering Strait: An Overview of Historic Sites in the BSNC Region*, vol. 1, *The Shishmaref Area* (Fairbanks, AK: Anthropology and Historic Preservation Cooperative Park Studies Unit, 1981).

28. Interview conducted by Elizabeth Marino with Fred Eningowuk in Shishmaref, Alaska, September 25, 2009.

29. Amanda J. Adler, Edward J. Boyko, Cynthia D. Schraer, and Neil J. Murphy, "Lower Prevalence of Impaired Glucose Tolerance and Diabetes Associated with Daily Seal Oil or Salmon Consumption Among Alaska Natives," *Diabetes Care* 17, no. 12 (1994): 1498–501; and Joseph Scanlon, "Winners and Losers: Some Thoughts about the Political Economy of Disaster," *International Journal of Mass Emergencies and Disasters* 6, no. 1 (1988): 47–63.

30. Kristin Dow and Thomas E. Downing, *The Atlas of Climate Change: Mapping the World's Greatest Challenge* (Berkeley: University of California Press, 2006), 37.

31. Candis L. Callison, "More Information Is Not the Problem: Spinning Climate Change, Vernaculars, and Emergent Forms of Life" (PhD dissertation, MIT, 2010), 11.

32. FEMA director Michael Brown on CNN, aired September 1, 2005.

33. http://en.wikipedia.org/wiki/Act_of_God.

34. Edward B. Barbier, "Upstream Dams and Downstream Water Allocation: The Case of the Hadejia-Jama'are Floodplain, Northern Nigeria," paper prepared for the Environmental Policy Forum, Center for Environmental Science and Policy, Institute for International Studies, Stanford University, November 7, 2002.

35. Interview conducted by Elizabeth Marino with Anonymous 1.a in Shishmaref, September 23, 2009.

36. Interview conducted by Elizabeth Marino with Jennifer Demir in Shishmaref, September 23, 2009.

37. Interview conducted by Elizabeth Marino and Stacey Stasenko, former Shishmaref resident, with Anonymous 3.b, September, 23, 2010.

38. Susan Hoffman and Anthony Oliver-Smith, "Introduction: Why Anthropologists Should Study Disasters," in *Culture & Catastrophe: The Anthropology of Disaster*, ed. Susanna Hoffman and Anthony Oliver-Smith (Santa Fe, NM: School of American Research Press, 2002), 4.

39. John E. Butler, *Natural Disasters* (Richmond: Heinemann Educational Australia, 1976), 69–70.

40. Harold Brooks, "Mobile Home Tornado Fatalities: Some Observations," NOAA/ERL/ National Severe Storms Laboratory, *www.nssl.noaa.gov/users/brooks/public_html/essays/ mobilehome.html*, accessed February 27, 2015.

41. For reviews and further information, see Kenneth Hewitt, *Interpretation of Calamity: From the Viewpoint of Human Ecology* (Boston: Allen and Unwin, 1983); and Anthony Oliver-Smith, "Anthropological Research on Hazards and Disasters," *Annual Review of Anthropology* 25 (1996): 303–28.

42. Hans-Martin Füssel, "Vulnerability: A Generally Applicable Conceptual Framework for Climate Change Research," *Global Environmental Change* 17 (2007): 155.

43. For reviews, see ibid.; Elizabeth Marino and Jesse Ribot, "Special Issue Introduction: Adding Insult to Injury: Climate Change, Social Stratification, and the Inequities of Intervention," in "Adding Insult to Injury: Climate Change, Social Stratification, and the Inequities of Intervention," special issue, *Global Environmental Change* 22 (2012): 323–28; Jesse Ribot, "Vulnerability Does Not Fall from the Sky," in *Social Dimensions of Climate Change: Equity and Vulnerability in a Warming World*, ed. Robin Mearns and Andrew Norton (Washington, DC: The World Bank, 2010), 47–74; Karen O'Brien, Siri Eriksen, Lynn P. Nygaard, and Ane Schjolden, "Why Different Interpretations of Vulnerability Matter in Climate

Change Discourses," *Climate Policy* 7, no. 1 (2007): 73–88; W. Neil Adger, "Vulnerability," *Global Environmental Change* 16, no. 3 (2006): 268–81; T. E. Downing, R. Butterfield, S. Cohen, S. Huq, R. Moss, A. Rahman, Y. Sokona, and L. Stephen, *Climate Change Vulnerability: Linking Impacts and Adaptation, Report to the Governing Council of the United Nations Programme* (Nairobi: United Nations Environmental Programme, 2001).

44. S. Schneiderbauer and D. Ehrlich, "Risk, Hazard and People's Vulnerability to Natural Hazards—A Review of Definitions, Concepts and Data" (Luxembourg: European Commission, 2004), 13.

45. Adger, "Vulnerability."

46. Based on Amartya Sen, *Resources, Values and Development* (Oxford: Blackwell, 1984).

47. Adger, "Vulnerability," 261.

48. For a review, see Hewitt, *Interpretation of Calamity*; Maxx Dilley and Tanya E. Boudreau, "Coming to Terms with Vulnerability: A Critique of the Food Security Definition," *Food Policy* 26, no. 3 (2001): 229–47; Susan Cutter, "Vulnerability to Environmental Hazards," *Progress in Human Geography* 20, no. 4 (1996): 529–39; Susan Cutter, "GI Science, Disasters, and Emergency Management," *Transactions in GIS* 7, no. 4 (2003): 439–45; Susan Cutter, "The Geography of Social Vulnerability: Race, Class, and Catastrophe," *Understanding Katrina: Perspectives from the Social Sciences,* accessed October 13, 2009, http://understandingkatrina.ssrc.org/Cutter; Susan Cutter, Lindsey Barnes, Melissa Berry, Christopher Burton, Elijah Evans, Eric Tate, and Jennifer Webb, "A Place-Based Model for Understanding Community Resilience to Natural Disasters," *Global Environmental Change* 18, no. 4 (2008): 598–606.

49. Cutter, "The Geography of Social Vulnerability."

50. Ben Wisner, Piers Blaikie, Terry Cannon, and Ian Davis, *At Risk: Natural Hazards, People's Vulnerability and Disasters,* 2nd ed. (New York: Routledge, 2003), 49.

51. For a review, see Michael J. Watts and Hans G. Bohle, "The Space of Vulnerability: The Causal Structure of Hunter and Famine," *Progress in Human Geography* 17, no. 1 (1993): 43–67; Oliver-Smith, "Anthropological Research on Hazards and Disasters"; David S. G. Thomas and Chasca Twyman, "Equity and Justice in Climate Change Adaptation Amongst Natural-Resource-Dependent Societies," *Global Environmental Change* 15 (2005): 115–24; Susan Cutter and Christopher Emrich, "Moral Hazard, Social Catastrophe: The Changing Face of Vulnerability Along the Hurricane Coasts," *Annals of the American Academy of Political and Social Science* 604 (2006): 102–12.

52. ACIA, *Arctic Climate Impact Assessment* (Cambridge: Cambridge University Press, 2005).

53. IPCC, "Policymakers' Summary," in *Climate Change: The IPCC Impacts Assessment,* ed. W. J. McG. Tegart, G. W. Sheldon, and D. C. Griffiths (Canberra: Australian Government Publishing Service, 1990); IPCC, "Summary for Policymakers," in *IPCC Special Report on Renewable Energy Sources and Climate Change Mitigation,* ed. O. Edenhofer, R. Pichs-Madruga, Y. Sokona, K. Seyboth, P. Matschoss, S. Kadner, T. Zwickel, et al. (Cambridge: Cambridge University Press, 2011).

54. G. McGranahan, D. Balk, and B. Anderson, "The Rising Tide: Assessing the Risks of Climate Change and Human Settlements in Low Elevation Coastal Zones," *Environment and Urbanization* 19, no. 1 (2007): 17–37.

55. Richard Moritz, Cecilia M. Bitz, and Eric J. Steig, "Dynamics of Recent Climate Change in the Arctic," *Science* 297 (2002): 1497–1502; Hinzman et al., "Evidence and Implications," ACIA, *Arctic Climate Impact Assessment*.

56. Peter Larsen, Scott Goldsmith, Orson Smith, Meghan L. Wilson, Ken Strzepek, Paul Chinowsky, and Ben Saylor, "Estimating Future Costs for Alaska Public Infrastructure at Risk from Climate Change," *Global Environmental Change* 18, no. 3 (2008): 442–57, doi:10.1016/j.gloenvcha.2008.03.005.

57. H. J. Fowler, S. Blenkinsop, and C. Tebaldi, "Linking Climate Change Modelling to Impacts Studies: Recent Advances in Downscaling Techniques for Hydrological Modelling," *International Journal of Climatology* 27 (2007): 1547–78.

58. Hinzman et al., "Evidence and Implications."

59. Igor Krupnik, and Dyanna Jolly, "The Earth Is Faster Now: Indigenous Observations of Arctic Environmental Change," Frontiers in Polar Social Science (Fairbanks, Alaska: Arctic Research Consortium of the United States, 2002).

60. J. J. Simpson, G. L. Hufford, M. D. Fleming, J. S. Berg, & J. B. Ashton, "Long-term Climate Patterns in Alaskan Surface Temperature and Precipitation and Their Biological Consequences," *IEEE Transactions on Geoscience and Remote Sensing*, 40 (2002): 1164–84.

61. T. E. Osterkamp and V. E. Romanovsky, "Evidence for Warming and Thawing of Discontinuous Permafrost in Alaska," *Permafrost Periglacial Process* 10 (1999): 17–37.

62. G. D. Clow and F. E. Urban, "Large Permafrost Warming in Northern Alaska During the 1990's Determined from GTN-P Borehole Temperature Measurements," 83, no. 47 (2002), Suppl., Abstract B11E-04, Fall Meeting, Eos Trans. American Geophysical Union, December 6–10, San Francisco.

63. V. E. Romanovsky, M. Burgess, S. Smith, K. Yoshikawa, and J. Brown, "Permafrost Temperature Records: Indicators of Climate Change," *Eos* 83, no. 50 (2002): 589/593–594.

64. Kenji Yoshikawa and Larry D. Hinzman, "Shrinking Thermokarst Ponds and Groundwater Dynamics in Discontinuous Permafrost near Council, Alaska," *Permafrost and Periglacial Processes* 14 (2003): 152.

65. Ibid., 151–60.

66. J. Magnuson, D. Robertson, B. Benson, R. Wynne, D. Livingstone, T. Arai, R. Assel, R. Barry, V. Card, E. Kuusisto, N. Granin, T. Prowse, K. Steward, and V. Vuglinski, "Historical Trends in Lake and River ice Cover in the northern Hemisphere," Science 289 (2000): 1743–46.

67. K. Ruhland, A. Priesnitz, and J. P. Smol, "Paleolimnological Evidence from Diatoms for recent Environmental Changes in 50 Lakes across Canadian Arctic Treeline," *Arctic, Antarctic, Alpine Res.* 35, no. 1 (2003): 110–23.

68. T. E. Osterkamp, L. Viereck, Y. Shur, M. T. Jorgenson, C. H. Racine, A. P. Doyle, and R. D. Boone, "Observations of Thermokarst in Boreal Forests in Alaska," *Arctic, Antarctic, Alpine Res.* 32, no. 3 (2000): 303–15.

69. Callison, "More Information Is Not the Problem." 55.

70. Elizabeth Marino and Peter Schweitzer, "Talking and Not Talking About Climate Change," in *Anthropology and Climate Change*, ed. Susan Crate and Mark Nuttal (Walnut Creek, CA: Left Coast Press, 2009), 209–17.

71. In the fall of 2011 the Seward Peninsula braced for a large storm that threatened most villages on the Peninsula with flooding. The storm did cause damage to fishing racks and other infrastructure in Shishmaref, but the prevailing winds were not from the southeast, making the storm surge much less impactful than it may have been if the winds had shifted.

72. Owen K. Mason, et al. "Narratives of Shoreline Erosion and Protection at Shishmaref, Alaska: The Anecdotal and the Analytical." *Pitfalls of Shoreline Stabilization* (2012): 73–92.

73. Owen Mason, William Neal, Orrin Pilkey, Jane Bullock, Ted Fathauer, Deborah Pilkey, and Douglas Swanston, *Living with the Coast of Alaska* (Durham, NC: Duke University Press, 1997).

74. Yoshikawa and Hinzman, "Shrinking Thermokarst Ponds," 151–60; and Molly Chambers, Daniel White, Robert Busey, Larry Hinzman, Lilian Alessa, and Andrew Kliskey, "Potential Impacts of a Changing Arctic on Community Water Sources on the Seward Peninsula, Alaska," *Journal of Geophysical Research* 112 (2007): 2.

75. Mason et al., *Living with the Coast of Alaska*, 106–10; Mason et al., *Narratives of Shoreline Erosion*.

76. Mason et al., *Narratives of Shoreline Erosion*.

77. USACE, *Alaska Village Erosion Technical Assistance Program*, 32.

78. Interview with Fred Eningowuk.

79. Interview conducted by Elizabeth Marino with Tony Weyiouanna in Shishmaref, Alaska, July 2008.

80. Interview conducted by Elizabeth Marino with John Sinnok in Shishmaref, Alaska, July 18, 2008.

81. Interview with Fred Eningowuk.

82. Interview conducted by Elizabeth Marino with Tommy Obruk in Shishmaref, Alaska, May 17, 2010.

83. Interview conducted by Elizabeth Marino with Tony Weyiouanna in Shishmaref, Alaska, July 2008.

84. Interview with Fred Eningowuk.

85. Bruce Lutz, "Population Pressure and Climate as Dynamics within the Arctic Small Tool Tradition of Alaska," *Arctic Anthropology* 19, no. 2 (1982): 143; J. L. Giddings, "The Archeology of Bering Strait," *Current Anthropology* 1, no. 2 (1960): 122.

86. Giddings, "The Archeology of Bering Strait," 121.

87. Josh Wisniewski, "Come On Ugzruk, Let Me Win: Experience, Relationality and Knowing in Kigiqtaamiut Hunting and Ethnography" (master's thesis, University of Alaska Fairbanks, 2011), 46.

88. Burch, *Eskimo Kinsman*; Koutsky, *Early Days on Norton Sound*.

89. Ernest Burch, *Social Life in Northwest Alaska: The Structure of Iñupiaq Eskimo Nations* (Fairbanks: University of Alaska Press, 2006), 31–52.

90. Alternately identified as a "society" (Ibid., 1) or a "tribe," the Iñupiaq word for these family groups is *nunaqatigiitch*, "people related to each other through possession of the land" (Burch, *The Iñupiaq Eskimo Nations*, 14; Burch, *Social Life in Northwest Alaska*, 29).

91. Burch, *The Iñupiaq Eskimo Nations*.

92. Burch, *Social Life in Northwest Alaska*, 45.

93. Interview conducted by Elizabeth Marino with Clifford Weyiouanna in Shishmaref, Alaska, July 21, 2008.

94. Ibid., 7.

95. Mason et al., *Narratives of Shoreline Erosion*.

96. Ibid.

97. Susan Fair, "Iñupiat Naming and Community History: The Tapqaq and Aniniq Coasts Near Shishmaref, Alaska," *The Professional Geographer* 49, no. 4 (1997): 472.

98. Maintenance of traditional subsistence land tenure makes the relocation of Shishmaref residents into a neighboring village problematic. A primary finding from the Army Corps of Engineers' cultural-impact assessment regarding relocating Shishmaref [Peter Schweitzer and Elizabeth Marino, *Coastal Erosion Protection and Community Relocation Shishmaref, Alaska: Collocation Cultural Impact Assessment* (Seattle: TetraTech, Inc., 2006)] was that relocating to a nearby village was not a tenable solution for permanent relocation. Shishmaref residents reported instances of historical violence between the *Kigiqtaamiut* people and residents of some villages to the North. Residents also commented that they would not have access to berry patches and hunting areas, as these areas were delineated for people from the area.

99. Interview with Clifford Weyiouanna.

100. Mobility into another nation's territory was also an important social insurance during times of scarcity and food insecurity, though one that could be lethal without appropriate social alliances [Ernest S. Burch Jr. and Thomas C. Correll, "Alliance and Conflict: Inter-regional Relations in North Alaska," *Alliance in Eskimo Society* (1972): 17–39]. Schweitzer and Golovko write, "contacts among individuals from different communities were always potentially problematic and hostile, as long as no kinship or partnership relations had been established. Individuals who had such relationships in other communities could travel freely and thus extend their existing social networks" [Peter P. Schweitzer and Evgeniy V. Golovko, "Local Identities and traveling Names: Interethnic Aspects of personal Naming in the Bering Strait Area," *Arctic Anthropology* (1997): 175].

101. Interview with Tommy Obruk.

102. Anna Huseth, letter written between 1919 and 1928 (n.d.), accessed November 29, 2012, http://wp.stolaf.edu/archives/the-heroine-of-the-north-the-works-of-sister-anna-huseth/

103. Elmer W. Ekblaw, "The Material Response of the Polar Eskimo to Their Far Arctic Environment," *Annals of the Association of American Geographers* 17, no. 4 (1927): 147–98.

104. Adele Perry, "From the Hot-Bed of Vice to the Good and Well-Ordered Christian Home: First Nations Housing and Reform in the Nineteenth-Century British Columbia," *Ethnohistory* 50, no. 4 (2003): 587.

105. M. W. Grauman, *A Historical Overview of the Seward Peninsula-Kotzebue Sound Area* (Washington, DC: National Park Service, 1977), 13–14.

106. Dorothy Jean Ray, *The Eskimos of the Bering Strait 1650–1898* (Seattle: University of Washington Press, 1975), 57; Linda J. Ellanna and George K. Sherrod, *From Hunters to Herders: The Transformation of Earth, Society, and Heaven Among the Iñupiat of Beringia* (Anchorage, AK: National Park Service, 2004).

107. Wisniewski, "Come On Ugzruk, Let Me Win," 52.

108. Ellanna and Sherrod, *From Hunters to Herders*, 23.

109. Koutsky, *Early Days on Norton Sound*.

110. Ellanna and Sherrod, *From Hunters to Herders*, 6.

111. Gigi Berardi, "Schools, Settlement, and Sanitation in Alaska Native Villages," *Ethnohistory* 46, no. 2 (1999): 333–35.

112. Katherine Stewart, "New Anti-Science Assault on US Schools," *The Guardian*, *www .guardian.co.uk/commentisfree/cifamerica/2012/feb/12/new-anti-science-assault-us-schools*, accessed November 2, 2012, see p. 263; Ellanna and Sherrod, *From Hunters to Herders*, 73.

113. For support, see Burch, *The Iñupiaq Eskimo Nations*, 47–50; and Wisniewski, "Come On Ugzruk, Let Me Win," 60. For detractors, see Ellanna and Sherrod, *From Hunters to Herders*, 76.

114. Ellanna and Sherrod, *From Hunters to Herders*, 11.

115. For a discussion, see ibid., 153–83.

116. Wisniewski, "Come On Ugzruk, Let Me Win," 70.

117. Ellanna and Sherrod, *From Hunters to Herders*.

118. Bruce M. Botelho, memorandum to Senator Halford and Senator Sharp, A.G. file no: 661-960688, April 26, 1996, p. 3.

119. Thomas W. Hennessy, Troy Ritter, Robert C. Holman, Dana L. Bruden, Krista L. Yorita, Lisa Bulkow, James E. Cheek, Rosalyn J. Singleton, and Jeff Smith, "The Relationship Between In-Home Water Service and the Risk of Respiratory Tract, Skin, and Gastrointestinal Tract Infections Among Rural Alaska Natives," *American Journal of Public Health* 98, no. 11 (2008): 2072–78.

120. Interview conducted by Elizabeth Marino with Anonymous in Shishmaref, Alaska, September 25, 2009.

121. Interview conducted by Elizabeth Marino with Anonymous in Shishmaref, Alaska, September 25, 2009.

122. USACE, *Alaska Village Erosion Technical Assistance Program*, 6.

123. Ibid., 32.

124. Interview with John Sinnok.

125. USACE, *Alaska Village Erosion Technical Assistance Program*, 32.

126. Interview with Jennifer Demir.

127. Owen Mason, "Living with the Coast of Alaska Revisited: The Good, the Bad, and the Ugly," in *Coastal Erosion Responses for Alaska*, ed. Orson P. Smith (Fairbanks: Alaska Sea Grant College Program, 2006), 11.

128. James Scott, *Seeing Like a State* (New Haven, CT: Yale University Press, 1998).

129. Ellanna and Sherrod, *From Hunters to Herders*, 11.

130. GAO, *Alaska Native Villages: Most Are Affected*.

131. Interview conducted by Elizabeth Marino with Richard Kuzuguk in Shishmaref, Alaska, September 24, 2009.

132. Division of Community and Regional Affairs (DCRA), *Background Information on the Shishmaref Relocation Effort* (Anchorage: DCRA, 1974), *www.commerce.state.ak.us/dcra/plans/Shishmaref-OT-1974.pdf*, accessed November 27, 2012.

133. Mason et al., *Living with the Coast of Alaska*.

134. Shishmaref Erosion and Relocation Coalition, "Shishmaref Strategic Relocation Plan," *www.google.com/search?client=safari&rls=en&q=shishmaref+strategic+relocation+plan&ie =UTF-8&oe=UTF-8*, accessed December 16, 2012.

135. Interview with Jennifer Demir.

136. Interview conducted by Elizabeth Marino with Richard Kuzuguk in Shishmaref, Alaska, September 24, 2009.

137. Immediate Action Working Group (IAWG), "Recommendations Report to the Governor's Subcabinet on Climate Change," *www.climatechange.alaska.gov/docs/iaw_rpt_17apr08.pdf*, p. 4, accessed December 9, 2011.

138. Joseph B. Verrengia, "In Alaska, an Ancestral Island Home Falls Victim to Global Warming," Associated Press, *http://forests.org/shared/reader/welcome.aspx?linkid=15600&keybold=climate%20AND%20%20seal%20AND%20%20level%20AND%20%20rise*, accessed March 1, 2015.

139. John D. Sutter, "Climate Change Threatens Life in Shishmaref, Alaska," CNN Tech: Climate Change, *www.cnn.com/2009/TECH/science/12/03/shishmaref.alaska.climate.change/index.html?eref=rss_tech*, accessed February 26, 2012.

140. Tommy Wallach, "The New Cold War: How Can a Tiny Island in Alaska's Chukchi Sea Beat Back Global Warming?" *Ready Made Magazine*, *www.readymade.com/magazine/article/the_new_cold_war*, accessed February 26, 2012. Since removed.

141. Carol Farbotko and Heather Lazrus, "The First Climate Refugees? Contesting Global Narratives of Climate Change in Tuvalu," in "Adding Insult to Injury: Climate Change, Social Stratification, and the Inequities of Intervention," special issue, *Global Environmental Change* 22 (2012): 382–90.

142. See *www.climatechange.alaska.gov/*, accessed May 10, 2010. Since removed.

143. Interview with Tommy Obruk.

144. Interview conducted by Elizabeth Marino with Kim and Stella Ningealook, May 13, 2010.

145. Interview conducted by Elizabeth Marino with Steve Sammons in Shishmaref, Alaska.

146. Interview with Jennifer Demir.

147. Interview with Anonymous 1.a.

148. Interview conducted by Elizabeth Marino with Daniel Iyatunguk in Shishmaref, Alaska, July 17, 2008.

149. Interview with Richard Kuzuguk.

150. Robin Bronen, "Climate-Induced Community Relocations: Creating an Adaptive Governance Framework Based in Human Rights Doctrine," *New York University Review of Law and Social Change* 35 (2011): 356–406.

151. See IAWG, "Recommendations Report"; Robin Bronen, "Forced Migration of Alaskan Indigenous Communities Due to Climate Change: Creating a Human Rights Response," in *Linking Environmental Change, Migration and Social Vulnerability*, ed. Anthony Oliver-Smith and Xiaomeng Shen (Bonn: UNU-EHS, 2009), 68–74, *www.ehs.unu.edu/file/get/4019*, accessed March 1, 2015; Bronen, "Climate-Induced Community Relocations"; AFN, "2009 Federal Priorities"; David Atkinson, Michael Black, Jessica Cherry, Billy Conner, et al., *Decision-making for At-Risk Communities in a Changing Climate* (Fairbanks, AK: Alaska Center for Climate Assessment and Policy, 2009).

152. For a more complete breakdown of policy obstacles to relocation, please see Bronen, "Climate-Induced Community Relocations." Dr. Bronen and I are in the process of researching the limits of the Voluntary Buyout Program (VBP) as a possible mechanism to facilitate environmental migration as a climate change adaptation strategy, but to date this is not what the VBP is equipped to handle.

153. Shihva Polefka, "Moving Out of Harm's Way," Center for American Progress, *www.americanprogress.org/issues/green/report/2013/12/12/81046/moving-out-of-harms-way/*, accessed April 30, 2014.

154. R. D. Knabb, J. R. Rhome, and D. P. Brown, "Tropical Cyclone Report," National Weather Service, *www.nhc.noaa.gov/pdf/TCR-AL122005_Katrina.pdf*, accessed April 30, 2014.

155. IAWG, "Recommendations Report to the Governor's Subcabinet on Climate Change," accessed December 9, 2011.

156. Bronen, "Forced Migration of Alaskan Indigenous Communities," 6.

157. Ibid.; personal observations.

158. Interview conducted by Elizabeth Marino with Brice Eningowuk in Shishmaref, Alaska, September 24, 2009.

159. Interview with Tommy Obruk.

160. Bronen, "Climate-Induced Community Relocations," 382–83.

161. Testimony of the Shishmaref Erosion and Relocation Coalition Before the Committee on Homeland Security and Governmental Affairs, Sub Committee on Disaster Recovery. United States Senate, October 11, 2007.

162. *www.tribesandclimatechange.org/docs/tribes_482.pdf*

163. The website: http://shishmarefrelocation.com/ has been taken down due, in part, to funding issues.

164. Sandra Sobelman, *The Economics of Wild Resource Use in Shishmaref, Alaska* (Juneau: Alaska Department of Fish and Game, Division of Subsistence, 1985), 4.

165. Annie Olanna Conger and James Magdanz, *The Harvest of Fish and Wildlife in Three Alaska Communities: Brevig Mission, Golovin, and Shishmaref* (Juneau: Alaska Department of Fish and Game, Division of Subsistence, 1990), 29.

166. Ibid., 27.

167. Thomas Thorton, "Alaska Native Subsistence: A Matter of Cultural Survival," *Cultural Survival* 22, no. 3 (1998), *www.culturalsurvival.org/ourpublications/csq/article/alaska-native-subsistence-a-matter-cultural-survival*, accessed November 28, 2012.

168. Interview with John Sinnok.

169. Interview conducted by Elizabeth Marino with Raymond Weyiouanna in Shishmaref, Alaska, July 16, 2008.

170. Phyllis Morrow, "Symbolic Actions, Indirect Expressions: Limits to Interpretations of Yupik Society," *Études/Inuit/Studies* 14, no. 1–2 (1990): 154; Deanna Paniataaq Kingson, "The Persistence of Conflict Avoidance Among the King Island Iñupiat," *Études/Inuit/Studies* 32, no. 2 (2008): 158.

171. Interview with Fred Eningowuk.

172. Interview with Raymond Weyiouanna.

173. Interview conducted by Elizabeth Marino with Esther Iyatunguk, May 21, 2010.

174. Interview with Brice Eningowuk.

175. Schweitzer and Marino, *Coastal Erosion Protection*, 67.

176. Clifford Geertz, "From the Native's Point of View: On the Natural of Anthropological Understanding," *Bulletin of the American Academy of Arts and Sciences* 28, no. 1 (1974): 31.

177. For a review, see A. P. Fiske, S. Kitayama, H. R. Markus, and R. E. Nisbett, "The Cultural Matrix of Social Psychology," in *Handbook of Social Psychology*, 4th ed., vol. 2, ed. D. T. Gilbert, S. Fiske, and G. Lindzey (New York: McGraw-Hill, 1998), 915–81; Patricia M. Greenfield, "Linking Social Change and Development Change: Shifting Pathways of Human Development," *Developmental Psychology* 45, no. 2 (2009): 401–18; H. C.

Triandis, "Collectivism and Individualism as Cultural Syndromes," *Cross-Cultural Research* 27 (1995): 155–80; Hazel Rose Markus and Shinobu Kitayama, "Culture and the Self: Implications for Cognition, Emotion, and Motivation," *Psychological Review* 98 (1991): 224–53.

178. Hazel Rose Markus and Shinobu Kitayama, "Cultures and Selves: A Cycle of Mutual Constitution," *Perspectives on Psychological Science* 5, no. 4 (2010): 420–30.

179. Tim Ingold, *The Appropriation of Nature: Essays on Human Ecology and Social Relations* (Manchester, UK: Manchester University Press, 1986), 184.

180. Paul Nadasdy, "The Gift of the Animal: The Ontology of Hunting and Human-Animal Sociality," *American Ethnologist* 34, no. 1 (2007): 25–43; Ingold, *The Appropriation of Nature*.

181. Tim Ingold, "Rethinking the Animate, Re-animating Thought," *Ethnos* 71, no. 1 (2006): 9–20.

182. Wisniewski, "Come On Ugzruk, Let Me Win," 141.

183. IAWG, "Recommendations Report to the Governor's Subcabinet on Climate Change," *www.climatechange.alaska.gov/docs/iaw_finalrpt_12mar09.pdf*, see p. 18, accessed December 9, 2011.

184. Ibid., 91.

185. Nadasdy, "The Gift of the Animal," 26.

186. Peter Schweitzer, Tobias Holzlehner, and Beth Mikow, "Beyond Forced Migration: A Typological Perspective on Resettlements in Alaska and Chukotka," in *Moved by the State: Population Movements and Agency in the Circumpolar North and Other Remote Regions*, ed. P. Schweitzer, E. Khlinovskaya-Rockhill, F. Stammler, and T. Heleniak (New York: Berghahn Books, forthcoming).

187. Chris De Wet, "Economic Development and Population Displacement: Can Everybody Win?" *Economic and Political Weekly* (2001): 4637–6.

188. Strickland, Rennard. "Absurd Ballet of American Indian Policy or American Indian Struggling with Ape on Tropical Landscape: An Afterword." *Me. L. Rev.* 31 (1979): 213.

189. Interview conducted by Elizabeth Marino with Clifford Weyiouanna in Shishmaref, Alaska, July 21, 2008.

190. Robert L. Kelly and Lawrence C. Todd, "Coming into the Country: Early Paleoindian Hunting and Mobility," *American Antiquity* 53, no. 2 (1988): 231–44; Jon Erlandson, Madonna Moss, and Matthew Des Lauriers, "Life on the Edge: Early Maritime Cultures of the Pacific Coast of North America," *Quaternary Science Reviews* 27 (2008): 2232–45.

191. For review, see Michael Cernea, "Risks, Safeguards, and Reconstruction: A Model for Population Displacement and Resettlement," in *Risks and Reconstruction: Experiences of Resettlers and Refugees*, ed. Michael Cernea and Chris McDowell (Washington DC: The World Bank, 2000), 11–55; Chris De Wet, "Risk, Complexity and Local Initiatives in Forced Resettlement Outcomes," in *Development-Induced Displacement*, ed. Chris De Wet (New York: Berhahn Books, 2006), 180–202; Anthony Oliver-Smith, "Disasters and Forced Migration in the 21st Century," *Understanding Katrina: Perspectives from the Social Sciences*, accessed May 8, 2011, http://understandingkatrina.ssrc.org/Oliver-Smith; Anthony Oliver-Smith, "Climate Change and Population Displacement: Disasters and Diasporas in the Twenty-First Century," in *Anthropology and Climate Change*, ed. Susan Crate and Mark Nuttall (Walnut Creek, CA: Left Coast Press, 2009), 116–38; Graeme Hugo, "Lessons from Past Forced Resettlement for Climate Change Migration," in *Migration and Climate*

Change, ed. Étienne Piguet, Antoine Pécoud, and Paul De Guchteneire (Cambridge: Cambridge University Press, 2011), 260–88.

192. Rennard Strickland, "The Absurd Ballet of American Indian Policy or American Indian Struggling with Ape on Tropical Landscape: An Afterward," *Maine Law Review* 31 (1979): 214.

193. Mildred E. Warner, "Civic Government or Market-Based Governance? The Limits of Privatization for Rural Local Governments," *Agriculture and Human Values* 26, no. 1–2 (2009): 133–43.

194. Tony Spybey, *Social Change, Development and Dependency: Modernity, Colonialism and the Development of the West* (Cambridge: Polity, 1992).

195. Matthew C. Snipp, "The Size and Distribution of the American Indian Population: Fertility, Mortality, Migration, and Residents," *Population Research and Policy Review* 16, no. 1–2 (1996): 66.

196. Schweitzer et al., "Beyond Forced Migration."

197. Callison, "More Information Is Not the Problem."

198. For a review, see Thomas and Twyman, "Equity and Justice"; Commission on Climate Change and Development, *Closing the Gaps: Disaster Risk Reduction and Adaptation to Climate Change in Developing Countries* (Stockholm: Commission on Climate Change and Development, 2009); Karen O'Brien and Robin Leichenko, "Double Exposure: Assessing the Impacts of Climate Change Within the Context of Economic Globalization," *Global Environmental Change* 10 (2000): 221–32; W. Neil Adger, Jouni Paavola, and Saleemul Huq, "Toward Justice in Adaptation to Climate Change," in *Fairness in Adaptation to Climate Change,* ed. W. Neil Adger, Jouni Paavola, Saleemul Huq, and M. J. Mace (Cambridge, MA: MIT Press, 2006), 1–19; Jesse Ribot, "Vulnerability Does Not Fall from the Sky," in *Social Dimensions of Climate Change: Equity and Vulnerability in a Warming World,* ed. Robin Mearns and Andrew Norton (Washington DC: The World Bank, 2010), 47–74.

199. Graeme Hugo, "Lessons from Past Forced Resettlement for Climate Change Migration," in *Migration and Climate Change,* ed. Étienne Piguet, Antoine Pécoud, and Paul De Guchteneire (Cambridge: Cambridge University Press, 2011), 260–88.

Index

ABC News, 63
ACCIMP (Alaska Climate Change Impact Mitigation Program), 66–67, 98
acculturation, 52
Act of God, 19, 20, 22
adaptability, 59, 85, 94–95. *See also* mobility
Adger, Neil, 24–25
"alarmist" literature, 10
Alaska. *See also specific locations, e.g.,* Shishmaref, AK
  climate change in, 5, 32–33
  colonialism and, 51–53. *See also* colonialism
  flooding in, 60, 63, 65
  indigenous peoples, 46–49. *See also* indigenous peoples
  statehood, 13
Alaska Climate Change Impact Mitigation Program (ACCIMP), 66–67, 98
Alaska Climate Change Sub-Cabinet, 65
Alaska District, Corps of Engineers, 42
Alaska Division of Homeland Security, 65
Alaska Municipal League, 65
Alaska Native Claims Settlement Act (ANCSA), 13, 96, 100
Alaska Native peoples/villages
  climate change and, 75–76, 95–96
  colonialism and, 51–53. *See also* colonialism
  flooding and erosion, 5, 60, 63, 91–92, 95–96
  global warming and, 9
  government policies and, 52, 59, 94, 95–96

housing assistance legislation, 54–55
interdependence of, 87–88
land claims acts, 13
planning coordination, 76–77
political ideology, 74, 77
relocation, 65–66, 70, 95–96. *See also* relocation
rights of, 92, 96–97, 99–100
service delivery and, 95
subsistence and, 81–92. *See also* subsistence
viability of, 95–96
Alaska State Legislative and Budget Committee, 65
Alaska Tribal Health Consortium, 65
albedo, 5
American Indians
  "civilization" programs, 52
  government policy and, 52, 59, 94, 95–96
  housing assistance programs, 54–55
  urbanization and, 95–96
ANCSA (Alaska Native Claims Settlement Act), 13, 96, 100
animals, culture and, 90, 91–92
animism, 90, 92
anjzugaksrat iniqtigutait, 90
anthropogenic climate change, 2, 3–4, 6, 9, 16, 24, 28, 35, 38–43, 58, 67, 96–97. *See also* climate change
anthropological tradition, 81–82, 88–89, 91–92, 97
apathy, 73

Arctic
    climate change, 4–5, 9, 16–17, 31, 32–33,
        35, 37–38, 85
    indigenous peoples. *See* indigenous peoples
    infrastructure, 16
    resettlement policies, 95
Arctic (Arctic River), AK, 77
Arctic amplification, 4–5, 32
Arctic Climate Impact Assessment, 31
Arctic Ocean, 5
Arctic Small Tool Tradition, 46
Army Corps of Engineers, 11, 41, 42, 56, 57,
        61, 62, 65, 81
Arrhenius, Svante, 3–4
artwork, 14
Associated Press, 63, 64
Australia, 4

Baltar, Julie, 77
Bangladesh, 7–8
barrier islands, characteristics of, 43, 57
BBC, 63
bearded seals, 15, 45
Beechy, Fredrick, 53
Bering Land Bridge, 46
Bering Strait, 2, 12, 14, 15, 46–47, 74
Bering Strait Native Corporation, 13
Bering Strait Superstorm, 40
BIA (Bureau of Indian Affairs), 43–44
Bournay, Emmanuelle, 7, 8
break-up, ice, 36–37
Brown, Michael, 19, 20
Burch, Ernest, 47
Bureau of Indian Affairs (BIA), 43–44
bureaucracy, 66–67, 69–70
Butler, John, 22

Cape Espenberg, 38, 49, 53
carbon dioxide, 4
Caribbean, 7–8
case study research, 93–94, 97
CBS News, 63
Central Asia, 7–8
Christianity, 52–53, 56–58
Christianization, 95

Chukchi Sea, 1, 5, 11, 12, 32
churches, 11, 13, 52–54
circle of subsistence, 13–14
citizen participation/activism, 71–74, 98
City of Shishmaref, 62. *See also*
        Shishmaref, AK
"civilization," 13, 52, 95
climate change
    in Alaska, 5, 32–33
    Alaska Natives and, 75–76, 95–96
    anthropogenic, 2, 3–4, 6, 9, 16, 24, 28, 35,
        38–43, 58, 67, 96–9
    Arctic and, 16–17, 31, 85
    in Arctic, 4–5, 9, 16–17, 31, 32–33, 35,
        37–38, 85
    debate, 4
    economic issues and, 70
    ethics and, 93–100
    ethnography of, 93–94
    human migration and, 6–10
    human rights and, 10
    media attention, 9, 10, 62–65
    modeling, 32–33
    natural disasters and, 28, 31–32
    pressure and release model and, 27–28
    research directions, 97–100
    in Shishmaref, 2–3, 16–17, 32–33, 38–43,
        69–70, 71–72
    vulnerability and, 31–32, 96–97
climate change refugees, 3, 4–10, 64,
        75–76, 94
Clow, G. D., 35
CNN, 9
coastal erosion. *See also* flooding and erosion
    anthropogenic, 39, 41–43
    in Shishmaref, 32, 37–38, 39, 41–43,
        56–58, 94
    shoreline stabilization and, 56–58, 94
    as slow-onset disaster, 31–32
coastal flooding. *See* flooding and erosion
colonialism
    in Alaska, 51–53
    Alaska Native peoples and, 51–53
    Christianity and, 52–53
    culpability and, 99–100

cultural imperialism/cultural values and, 16, 29, 58–60, 94, 95, 96
    "fourth world" challenges, 16
    indigenous peoples and, 16, 29, 51–53, 58–60
    infrastructure and, 51–53
    missionary ideology, 13, 45, 50, 51–53, 58
    racism and, 16, 29, 69–70, 73
    Seward Peninsula, 51–53
    Shishmaref and, 11, 16, 29, 44, 45–60
    subsistence territory and, 96, 99–100
    violence and, 51
Colorado University, 41
Conger, Annie Olanna, 82
Cox, Sally, 66–67
Creation Care, 3
Crosby, Thomas, 51
cross-agency relationships, 99
cross-cultural psychology, 88–89
cultural disintegration, 81–82, 94, 96, 97
cultural identity/values/traditions
    assimilation, 52
    colonialism and, 16, 29, 58–60, 94, 95, 96
    cross cultural psychology and, 88–89
    cultural disintegration, 81–82, 94, 96, 97
    home, tenacity of, 81–92
    infrastructure development and, 58–60
    relocation and, 81–92, 94, 96, 97
    subsistence and, 87–88, 90–91
cultural psychology, 88–89, 89–91
Cutter, Susan, 26, 28, 29
cyclones, 32

DCRA (Division of Community and Regional Affairs), 61–62, 66
De Wet, Chris, 93
Demir, Jennifer, 74
Denali Commission, 55, 65
Department of Commerce, Community, and Economic Development, 65, 66
Department of Housing and Urban Development (HUD), 54–55
Department of Natural Resources, 65
Department of Transportation and Public Facilities, 65

desertification, 6
development. *See* infrastructure/ infrastructure development
diaspora, 72, 78–79
diet, 15, 82–84
disaster response/prevention, 22–24, 71–74, 75–76, 97. *See also* risk management/ mitigation
disasters. *See* natural disasters
Division of Community and Regional Affairs (DCRA), 61–62, 66
drought, 2–3, 6, 9, 25–26, 31

earthquakes, 19–20, 32
Earthwatch Radio, 63
East Nunatuq, AK, 77
ecology
    ecological change, rapid pace of, 94–95
    ecological shifts, 94
    natural disasters effects and, 19–28, 32
    society and, 20–22
    socioecological systems, 78–79
    vulnerability and, 22–28, 32
economic issues. *See also* funding
    climate change and, 70
    economic intervention, 10
    human migration and, 6, 7
    natural disasters and, 19–20
education infrastructure, 13, 52, 53, 58–60, 45
El-Hinnawi, Essam, 6
Ellanna, Linda J., 54
Eningowuk, Brice, 88
Eningowuk, Fred, 13, 34, 43, 44, 45–46, 85, 87
entitlements, 24–26
entitlements approach, 25–26
environment. *See also* climate change; ecology
    interdependence and, 89–91
    vulnerability and, 24–25. *See also* vulnerability
environmental activism, 3
environmental justice, 30
environmental migrants/migration, 3, 4–10, 64, 75–76, 94
    hotspots, 7–8

Environmental Protection Agency, 65
environmental refugees, 3, 4–10, 63,
    75–76, 94
erosion. *See* flooding and erosion
"Eskimo law," 90
Eskimo peoples, 10, 85
ethics
    climate change and, 93–100
    vulnerability and, 96–97
Ethiopia, 9
ethnography, 11, 69, 93–94, 97
Europe, 7, 46
Evangelical environmentalism, 3
Evans, Peter, 100

Facebook, 86
Fair, Susan, 48
famine, 9, 25–26
Farbotko, Carol, 64
fatalities, 19–20, 23
Federal Emergency Management Agency
    (FEMA), 19, 20, 23, 75–76, 97
federal funding. *See* funding
FEMA (Federal Emergency Management
    Agency), 19, 20, 23, 75–76, 97
flexibility, mobility and, 50–51, 57–58
Flood Mitigation Assistance (FMA)
    program, 75
flooding and erosion
    in Alaska, 60, 63, 65
    Alaska Native villages, 5, 60, 63, 91–92,
        95–96
    anthropogenic, 39, 41–43
    buyout programs, 76
    climate change and, 31–32
    media attention, 65
    mobility and, 49–51
    prevention, 22–24
    in Shishmaref, 2–3, 7–9, 15–16, 21–22,
        31–44, 49–51, 56–68, 94
    as slow-onset diaster, 31–32
    society and ecology and, 20–22
    worldwide, 7–9
FMA (Flood Mitigation Assistance)
    program, 75

food insecurity, 19, 25–26
fossil fuels, 4
"fourth world," 16
freeze-up, 36–37
French Daily Liberation, 63
fresh water availability, 34–35
funding
    colonial era, 52–53, 95
    entitlements approach and, 26
    home ownership programs, 54–55
    infrastructure development/rebuilding,
        15–16, 26–27, 55, 76, 95
    relocation programs, 16, 62–63, 65, 66, 70,
        72, 75–76, 77, 78, 99–100
    subsistence territory, 99–100
Füssel, Hans-Martin, 24

Geertz, Clifford, 88
genocide, 51, 100
greenhouse gases, 9–10, 32–33, 39, 96–97
Giddings, J. L., 84, 85, 86
Gift of the Animal, The, 91
global citizenry, 64
Global Create (Japan), 63
global warming, 4–5, 9, 16–17, 28, 35,
    37–38, 39, 43, 70, 71–72, 93, 96
gold, 13
government
    American Indian/Alaska Natives policies,
        52, 59, 94, 95–96
    bureaucracy, 66–67, 69–70
    colonialism and, 51–53. *See also*
        colonialism
    culpability, 99–100
    distrust of, 69–70
    funding. *See* funding
    home ownership programs, 54–55
    land claim acts, 13
    natural disasters/disaster prevention
        policies, 22–24, 71–74, 75–76, 97
    relocation programs/policies, 62–66,
        75–76, 94, 95–96, 98, 99–100
    resettlement policies, 24, 93, 94, 95
    revolving-door theory and, 66–67, 98
    subsidization, 14, 95

greenhouse effect, 4
greenhouse gases, 3–4
Greenlandic Ice Sheet, 31

Hadejia-Jama'are floodplain, 21
Haiti, 7–8, 19–20, 32
Hamza, Mo, 20
Hazard Mitigation Grant Program
    (HMGP), 75
hazard mitigation policy/strategy, 97. *See
    also* disaster response/prevention; risk
    management/mitigation
hazard-centric policies, 22–24
hazards, 22. *See also* natural disasters
HBO, 63
HD Net TV, 63
health, 15, 55
Hinzman, Larry, 33, 34, 35, 36, 37
Hoffman, Susanna, 22
home, tenacity of, 81–92
home ownership programs, 54–55
Hopson, Rainey, 87, 89
hotspots, environmental migration, 7–8
housing, traditional, 50–51, 53–55,
    94–95
HUD (Department of Housing and Urban
    Development), 54–55
human migration
    climate change and, 6–10
    ecological shifts and, 94
    economics and, 6, 7
    environment and, 6–10
    environmental, 3, 4–10, 64, 75–76, 94
    mass, 7, 10
    patterns of, 10
    technology and, 6
human rights, 10, 11, 64
hunting, subsistence, 11–12, 13–14, 43–44,
    45–46, 71, 78
Hurricane Katrina, 20, 22, 26, 29, 76
Hurricane Rita, 22
Hurricane Sandy, 76
hurricanes, 20, 22, 26, 29, 32, 76
Huseth, Sister Anna, 50
hydrological systems, 34–35, 37

IAWG (Immediate Action Working Group),
    63, 65, 66, 67–69, 73, 76, 90–91
Igloot, Tin Creek, AK, 77
Ikpik village, 49, 53
Immediate Action Working Group
    (IAWG), 63, 65, 66, 67–69, 73, 76,
    90–91
independence, vs. interdependence, 89
India, 7–8
indigenous peoples. *See also* specific groups
    adaptability of, 59, 85, 94–95
    in Alaska, 46–49
    anthropological tradition and, 91–92
    artwork, 14
    "civilization" of, 13, 52, 95
    colonialism and, 16, 29, 51–53, 58–60. *See
        also* colonialism
    culture and traditions of, 81–92. *See also*
        cultural identity/values/traditions
    flexibility of, 50–51, 57–58
    mobility and. *See* mobility
    modernization and, 16, 19–20, 95
    in North America, 46–49
    racism and, 29–30, 73
    relocation and, 81–92, 95–96. *See also*
        relocation
    in Seward Peninsula, 13
    in Shishmaref, 9–10, 12–17, 46–49
    subsistence and, 81–92, 99–100
    subsistence territory, 11, 45–46, 81–82,
        85, 96–97, 99–100
Industrial Revolution, 4
industrialization, 9–10, 52
inequity, 96
infrastructure/infrastructure development
    Arctic, 16
    climate change and, 5
    colonialism and, 51–53
    cultural values and, 58–60
    education, 13, 45, 52, 53, 58–60
    flexibility and, 57–58
    funding, 15–16, 26–27, 55, 76, 95
    mobility and, 58–60, 94–95
    natural disasters and, 19–20
    relocation and, 55–56, 59, 78–79

infrastructure/infrastructure
    development (*continued*)
  Shishmaref and, 15–16, 50–51, 53–55,
    94–95
  shoreline stabilization and, 56–58
  tax base and, 26–27
  traditional housing, 50–51, 53–55, 94–95
  transportation, 12
  Western vs. traditional, 45
Ingold, Tim, 89–90
interdependence, 87–88, 89–91
Intergovernmental Panel on Climate Change
    (IPCC), 6, 24
International Organization on Migration
    (IOM), 6
Iñupiat/Iñupiaq peoples, 9–10, 12–17,
    46–49, 51, 53, 59, 78, 81–82, 85, 86,
    87–88
Iyatunguk, Esther, 86, 89
Izmit, Turkey, earthquake, 19–20

Jackson, Sheldon, 45, 52–53, 59, 73, 95
jewelry, 14
Jolly, Dyanna, 34

Kahlook Barr Sr., Gideon, 54
Kawerak nonprofit organization, 13
Kigiqtaamiut peoples, 12–17, 30, 43–44, 46,
    48–49, 51, 56, 78, 85, 90, 91, 93–94,
    95, 97
Kigitaq, 48, 51, 53, 54, 59
Kivalina, AK, 7–9, 66
Knowles, Tony, 62
Kokeok, Annie, 73, 87
Kokeok, Kate, 77
Kotzebue, AK, 11, 65, 88, 89
Koyukuk, AK, 7–9, 66
Krupnik, Igor, 34
kunituks, 15
Kuzuguk, Richard, 61
Kyoto Protocol, 4

Lake Charles, LA, 22
land claim acts, 13
landscape, interdependence and, 89–91

Lazrus, Heather, 64
Le Monde Diplomatique, 7, 8
Little Diomede, 74
local participation, 98
Loma Prieta earthquake, 19–20

Magdanz, James, 82
Magnunson, J., 37
Maison Radio (Canada), 63
marginalization, 20, 26–27, 28, 94, 96
Mason, Owen, 41–43, 48, 56, 58
mass migration, 7, 10
media attention, 9, 10, 62–65
Mekong River delta, 7–8
Mexico, 7–8
missionary ideology, 13, 45, 50, 51–53, 58
Missouri, 76
mobility
  adaptation and, 47–48, 49–51, 58–60, 78
  infrastructure and, 58–60, 94–95
  modernization, 16, 19–20, 95
Myers, Norman, 6, 7

Nadasdy, Paul, 91
NAHASDA (Native American Housing
    Assistance and Self-Determination
    Act), 55
National Film Board of Canada, 63
National Geographic magazine, 63
National Oceanic and Atmospheric
    Administration, 65
National Public Radio, 63
National Weather Service, 40
Native American Housing Assistance and Self-
    Determination Act (NAHASDA), 55
Native American peoples, 16, 95–96, 96–97.
    See also American Indians
Native Village of Shishmaref, 62. *See also*
    Shishmaref, AK
Nayokpuk, Herbert, 1
Nayokpuk, Percy, 62
natural disasters
  anthropogenic, 2–3
  climate change and, 31–32
  defined, 19–20

governmental policy, 22–24, 71–74, 75–76, 97
    hazard-centric vs. vulnerability, 22–24
    poverty, and 19–20, 24–25, 26–27, 28
    research directions, 97–100
    risk management, 71–74
    slow-onset vs. rapid-onset, 31–32
    society and ecology and, 20–22
    stigmatization and vulnerability and, 29–30
    vulnerability and, 24–29, 31–32, 96–97
New Orleans, LA, 7–8, 20, 22, 26
New York Times, 9, 63
New Yorker, 63
Newtok, AK, 7–9, 66, 76–77
Newtok Traditional Council, 76, 77
Nigeria, 21
Nile River, 21, 51
Ningealook, Kim, 71
Ningealook, Stella, 71
Nome, AK, 11, 13, 65, 88, 89
Nome Nugget, 22
North America, 6, 46–49. *See also* United States
Norwegian Broadcasting Corporation, 63
nutrition, 15

O'Brien, Erik, 66–67
Obruk, Tommy, 44, 50, 66
Oklahoma City, OK, 23
Old Shishmaref, 48, 59
Oliver-Smith, Anthony, 22
Oregon State University, 3
Osterkamp, T. E., 35, 38
outmigration, 56
overcrowding, 55–56
Oxford Brooks University, 20

Palin, Sarah, 65
panaaluk, 15
People magazine, 63
permafrost thaw, 34–35
Perry, Adele, 51
Phoenix, AZ, 6
polar amplification, 4–5, 32

political ecological model of disaster, 26–27
political ideology, 74
Port-au-Prince, Haiti, 19–20
poverty, 19–20, 24–25, 26–27, 28
Pre-Disaster Mitigation (PDM) program, 75
pressure and release model, 27–28
Pueblo Grande, 6
racism
    colonialism and, 16, 29, 69–70, 73
    indigenous peoples and, 29–30, 73
    vulnerability and, 20, 29–30
rapid-onset natural disasters, 31–32
rebuilding in place, 23, 97
relocation
    Alaska Native villages, 65–66, 70, 95–96
    cost comparison, 100
    cultural and traditions and, 81–92, 94, 96, 97
    funding, 16, 62–63, 65, 66, 70, 72, 75–76, 77, 78, 99–100
    government/government policies and, 61–62, 65–66, 66–67, 69–70, 71–74, 75–76
    indigenous peoples and, 81–92, 95–96
    individual households, 75–76, 97
    infrastructure and, 78–79
    need for central agency, 98
    research, 65
    revolving-door theory, 66–67
    site selection, 77–78, 99
    subsistence and, 81–92
relocation, Shishmaref
    colonialism and, 16
    contemporary discussion, 78–79
    cost comparison of, 100
    cultural and traditions and, 81–92, 94, 96, 97
    current state of affairs, 67
    funding, 16, 62–63, 65, 66, 70, 72, 75–76, 77, 78, 99–100
    infrastructure and, 55–56, 59
    management gaps, 71–74
    Netwok Planning Group, 76–77
    obstacles to, 69–70
    overview/summary, 11, 94

relocation, Shishmaref (*continued*)
  planning timeline, 61–66
  policy challenges, 75–76
  pragmatics, 67–69
  revolving-door theory, 66–67
  site selection, 77–78, 99
  subsistence and, 45–46
  tenacity of home and, 81–92
research
  anthropological tradition, 81–82, 88–89,
    91–92, 97
  bias, 6
  case study, 93–94, 97
  climate change modeling, 32–33
  erosion, 41–43
  ethnographic, 11, 69, 93–94, 97
  future directions, 97–100
  natural disasters, 22–24
resettlement policies, 24, 93, 94, 95
resiliency, 29–30, 71, 74, 96, 98
resources
  colonialism and, 16, 95, 100
  cyclical, 14, 49
  natural disasters and, 6, 20, 24–25, 26–27
  subsistence and, 82–83
  vulnerability and, 30, 93
Reuters, 63
revolving-door theory, 66–67
risk management/mitigation
  hazard-centric vs. vulnerability, 22–24
  pressure and release model, 27–28
  relocation programs and, 71–74
  subsistence and, 84–87
Romanovsky, V. E., 35
Rühland, K., 37
rural living, 87–88, 95
Russian Cossacks, 46

Sahel Belt, 7–8, 9
Samuels, Steve, 71
San Francisco Bay area, 19–20
Sarichef Island, 13, 33, 38, 42, 45–46, 51
sculpture, 14
scurvy, 15
sea-ice levels, 15–16

sea-level rise, 2–3, 6, 15–16, 31–32
seal oil, 15, 78
seals, 15, 45, 57, 83–84. *See also* subsistence
seawalls, 40, 41–43, 56–58
sedentarization, 50, 53, 58
self-determination, 98
Severe Repetitive Loss (SRL) grant
  program, 75
Seward Peninsula, 11, 12, 13, 15, 35–36,
  46–49, 50, 51–53, 59, 85, 86, 94
sewn crafts, 14
sexism, 52–53
Shaktoolik, 66
sharing customs, 83
Sherrod, George K., 54
Shishmaref
  adaptability, 94–95
  case study summary, 93–94
  climate change and, 2–3, 16–17, 32–33,
    38–43, 69–70, 71–72
  climate debate and, 3–4
  coastal erosion, 37–38, 39, 41–43, 56–58, 94
  colonialism and, 11, 16, 29, 44, 45–60
  community residents, valuing, 91–92
  cultural disintegration, 97
  demographics, 11
  diversity, 15
  drought, 2–3
  ecological and social change, 94–95
  economic sector, 14
  employment and, 14
  environmental migration, 3, 4–10
  ethics and, 96–97
  ethnographic fieldwork, 10–17
  flooding and erosion, 2–3, 7–9, 15–16,
    21–22, 31–44, 49–51, 56–68, 94
  food, 15
  freeze-up, 36–37
  government development, 52–53
  historical overview, 10–17
  housing, 53–55
  hunting, subsistence, 11–12, 13–14,
    43–44, 45–46, 71, 78
  indigenous peoples, 9–10, 12–17, 46–49.
    See also indigenous peoples

infrastructure, 15–16, 50–51, 53–55, 94–95

media attention, 9, 10, 62–65

mobility/adaptation and, 47–48, 49–51, 58–60, 78, 94–95

natural disasters and, 29–30

overcrowding, 55–56

overview, 1–2

permafrost thaw, 34–35

political organization, 13

population, 11

poverty and, 14

relocation of. See relocation, Shishmaref

research attention, 5–10

Sarichef Island as center, 45–46

sea-ice levels/sea-level rise, 15–16

shoreline stabilization, 56–58, 94

storm events, 15–16, 32, 56–57, 61–63

subsistence and, 11–12, 13–15, 43–44, 44–45, 71, 78, 81–88, 91–92, 94

thermokarst ponds, 35–36

transportation infrastructure, 12

vulnerability and, 26, 27, 29–30, 70

weather patterns, 33–34

Shishmaref Erosion and Relocation Coalition, 49, 62, 63, 76, 82

Shishmaref Erosion and Relocation Committee, 72

Shishmaref Inlet, 1, 11, 12, 51

Shishmaref Lutheran Church, 11, 53

Shishmaref Native Corporation, 62

shoreline stabilization, 56–58, 94

Siberia, 5

Siberian Native populations, 95

Simpson, J. J., 34

Sinnok, John, 57, 83

slow-onset natural disasters, 31–32

Sobelman, 82

social capital, 25

social change, rapid pace of, 94–95

social engineering, 73, 94

social justice, 10, 30, 96, 100

social media, 86

socioecological systems, 78–79

natural disasters effects and, 19–28, 32

vulnerability and, 22–28, 32

SRL (Severe Repetitive Loss) grant program, 75

Stafford Act, 75

Stasenko, Rachel, 11

Stasenko, Rich, 11

Stevens, Ted, 74

stigmatization, vulnerability and, 29–30

storm events, 9, 15–16, 32, 56–57, 61–63

storm shelters, 23, 24

storm surges, 32

Strickland, Rennard, 93, 94

sub-Saharan Africa, 7

subsidization, government, 95

subsistence

circle of, 13–14

culture and tradition and, 87–88, 90–91

hunting, 11–12, 13–14, 43–44, 45–46, 71, 78

indigenous peoples and, 81–92, 99–100

practices, 82–84, 84–87, 87–88

risk management and, 84–87

Shishmaref and, 11–12, 13–15, 43–44, 44–45, 71, 78, 81–88, 91–92, 94

territory, 11, 45–46, 81–82, 85, 96–97, 99–100

sustainability, vulnerability and, 28

Svenska Dagbladet (Sweden), 63

systems of inequity, 96

Tapqagmiut, 46–47, 50, 51, 56, 58, 81

tax base, 26–27

technology, 6

Teller village, 50

tenacity of home, 81–92

textiles, 14

Thalassa (French television), 63

thermokarst ponds, 35–36

Thorton, Thomas, 83

Time magazine, 9, 63

Tom, Stanley, 76, 77

top-down social engineering, 73

tornadoes, 23

traditional housing, 50–51, 53–55, 94–95
transportation infrastructure, 12
TV Asahi (Japan), 63

Ublasaun village, 54
ugruk, 15, 45, 57, 83–84
Unalakleet, relocation, 66
United Nations Environmental Programme
    (UNEP), 6, 7
United States. *See also* government
    climate change legislation, 4
    colonialism. *See* colonialism
    disaster response policy, 75–76, 97
    education and missionization goals, 52
    environmental migrants and, 7
    funding of relocation, 100
    housing development programs, 54–55
    land claim acts, 13
    research bias, 6
United States Housing Act, 54
University of Alaska Fairbanks, 86
Urban, F. E., 35
urbanization, 95–96
U.S. Army Corps of Engineers, 11, 41, 42,
    56, 57, 61, 62, 65, 81
U.S. Census, 11
U.S. Economic Development
    Administration, 65
U.S. Government Accounting Office
    (GAO), 64
U.S. Senate, 79

violence, colonialism and, 51
Viverra Films (Holland), 63
Voluntary Buyout Program, 75–76
von Kotzebue, Otto, 51
vulnerability
    climate change and, 31–32, 96–97
    conceptual diagram, 25
    defined, 24

entitlements approach, 25–26
ethical issues and, 96–97
hazard approach, 24–25
hazard-centric vs. 22–24
inequity and, 96
natural disasters, and 24–29, 31–32,
    78–79, 95–97
political ecological model of disaster, 26–27
pressure and release model, 27–28
racism and, 20, 29–30
research directions, 97–100
stigmatization and, 29–30

warning systems, 23
water and sewage systems, 55
water diversion, 6
wave action/wave energy, 42, 56
wealthy nations, human migration and, 6, 7
Weather Channel, 63
weather patterns, 33–34
West Tin Creek Flats, AK, 77
West Tin Creek Hills, AK, 77
Western European research bias, 6
Weyiouanna, Clifford, 29, 34, 36, 48, 49, 93,
    100, 88
Weyiouanna, Tony, 11, 39, 43, 45, 59, 63, 64,
    79, 100
whaling, 46
white flight, 26–27
Wikipedia, 7, 19
Wisner, Ben, 27
Wisniewski, Josh, 47, 90

xenophobia, 7

Yangtze River, 7–8
Yoshikawa, Kenji, 36
Yupiit, 85

ZDF, 63